Dietary and Non-Dietary Phytochemicals and Cancer

Special Issue Editor
Carmela Fimognari

MDPI

Special Issue Editor
Carmela Fimognari
University of Bologna
Italy

Editorial Office
MDPI AG
St. Alban-Anlage 66
Basel, Switzerland

This edition is a reprint of the Special Issue published online in the open access journal *Toxins* (ISSN 2072-6651) from 2016–2017 (available at: http://www.mdpi.com/journal/toxins/special_issues/phytochemicals_cancer).

For citation purposes, cite each article independently as indicated on the article page online and as indicated below:

Author 1; Author 2; Author 3 etc. Article title. *Journal Name*. **Year**. Article number/page range.

ISBN 978-3-03842-378-2 (Pbk)
ISBN 978-3-03842-379-9 (PDF)

Table of Contents

About the Guest Editor

Carmela Fimognari earned her PhD in Toxicology in 1997 from the University of Bologna, Italy. After training at the GSF-National Research Centre for Environment and Health, Munich, Germany, she focused her research on cancer chemoprevention and therapy by natural compounds and synthetic analogues with cytostatic and cell death-inducing potential. As an example, she and her co-workers investigated isothiocyanates and anthocyanins, known to exhibit anti-inflammatory and anti-cancer activities. She directs the Laboratory for Molecular and Cellular Toxicology at the Department for Life Quality Studies of the University of Bologna. She was appointed associate Professor of Toxicology at the Faculty of Pharmacy of the University of Bologna in 2014. She wrote over 90 publications on international journals and book chapters.

toxins

MDPI

Review

Introduction to the *Toxins* Special Issue on Dietary and Non-Dietary Phytochemicals and Cancer

Carmela Fimognari

Department for Life Quality Studies, Alma Mater Studiorum-University of Bologna, Corso d'Augusto 237, Rimini 47921, Italy; carmela.fimognari@unibo.it

Academic Editor: Nilgun E. Tumer
Received: 15 December 2016; Accepted: 21 December 2016; Published: 28 December 2016

The role of many phytochemicals in the modulation of the carcinogenesis process has been well documented by combining in vitro and animal studies, as well as epidemiological evidence. When acting in synergy, phytochemicals exert potential anti-cancer properties, and much progress has been made in defining their many biological activities at the molecular level. However, an interesting feature in the field of phytochemicals and cancer is the role of some phytochemicals in promoting cancer development. This Special Issue of *Toxins* aims to provide a comprehensive look at the contribution of dietary and non-dietary phytochemicals to cancer development and at the molecular mechanisms by which phytochemicals inhibit or promote cancer.

Cancer stem cells represent a small subset of tumor cells endowed with uncontrolled proliferative capacity and indefinite potential for self-renewal that drive tumorigenesis. Considering the potential of cancer stem cells in the initial development of cancer, resistance to therapy and metastasis, they have become a critical target for the identification and development of new approaches to fight cancer. Oh et al. present an overview of phytochemicals targeting signaling pathways involved in stemness maintenance and survival of cancer stem cells [1]. Some examples include cyclopamine from the corn lily, curcumin from turmeric, and piperine from black and long peppers, sulforaphane from cruciferous vegetables, the soy isoflavone genistein, and blueberry polyphenols. Lu et al. [2] report the effect of ovatodiolide—a macrocyclic diterpenoid compound isolated from *Anisomeles indica*—of blocking the self-renewal capability of breast cancer stem cells and downregulating the expression of stemness genes on human breast cancer stem cells.

The detailed review by Ismail et al. [3] focuses on ellagitannins, a class of phytochemicals widely investigated for their chemopreventive and anticancer activities. With the aim of delineating and predicting their actual clinical potential, the authors present the dietary sources, the pharmacokinetics, and the evidence on the chemopreventive efficacy and the anticancer activity of ellagitannins. The chemopreventive effects of ellagitannins are linked to their antioxidant and anti-inflammatory properties. Their anticancer activity is imputable to different mechanisms, including inhibition of pro-inflammatory pathway, cell-cycle arrest ability, and proapoptotic properties. The pleiotropic nature of the mechanisms behind the anticancer activity of ellegitannins is also demonstrated by their ability to block angiogenesis and inhibit endothelial cell growth. However, as the authors point out, orally administered ellagitannins are characterized by a limited bioavailability. Their widely recognized anticancer efficacy may lead to the adoption of different administration routes to overcome their low bioavailability, such as the intravenous route. However, their toxicological profile after intravenous administration is lacking.

This Special Issue also features several research papers reporting selective effects of phytochemicals against cancer cells, for which follow brief synopses.

The article by Farhan et al. [4] highlights mechanistic studies which provide insights into the cytotoxic and anticancer activity of catechins. Catechin, epicatechin, epigallocatechin, and epigallocatechin-3-gallate—the four major constituents of green tea—exert a prooxidant action through redox recycling of copper ions, and induce cellular DNA breakage in human peripheral lymphocytes. The presence of copper ions boosts their genotoxic effect. Catechins (especially epigallocatechin-3-gallate) have antiproliferative activity on breast cancer cells. Copper levels are much higher in cancer cells than in non-transformed cells. Thus, cancer cells would be more susceptible to redox cycling between copper ions and catechins to generate reactive oxygen species and thus DNA breakage. Accordingly, the authors demonstrate that normal breast epithelial cells are quite resistant to the treatment with catechins, but their culture in medium enriched with copper make them more susceptible to catechins' antiproliferative effect. Taking the results presented in the paper into account, catechins exert their preferential prooxidant cytotoxic effect against cancer cells. Chen et al. [5] investigate the effect of tenuifolide B—derived from the stems of *Cinnamomum tenuifolium*—on viability, cycle progression, apoptosis, reactive oxygen species production, mitochondrial depolarization, and DNA damage in transformed and non-transformed oral cells. Tenuifolide B reduces the viability of cancer cells through the induction of apoptotic cell death. The production of reactive oxygen species and the induction of DNA damage are probably involved in its cytotoxic effect. Of note, the effects of tenuifolide B seem to be selective for cancer cells. Indeed, its effects on normal oral cells are much less pronounced than on cancer cells. Nguyen et al. [6] show that papaya leaves are a potential source of anticancer compounds. They actually contain flavonoids or flavonoid glycosides, particularly compounds from the kaempferol and quercetin families. Moreover, aqueous and ethanolic extracts of leaves reduce the survival of human oral squamous cell carcinoma cells. Interestingly, the two fractions with acidic pH are selectively cytotoxic towards the cancer cells, and the phenolic and flavonoid content positively correlates with the differential effect on cell viability.

However, it is worth noting that some phytochemicals can have an ambivalent character (especially in the possible context of high dose therapeutic applications), and favor cancer development. Some papers of this Special Issue deal with this important aspect.

Turrini et al. [7] provide new insights into the anticancer mechanisms of the root extract of *Withania somnifera*, a plant used in Indian traditional medicine. The article presents the cytotoxic and cytostatic effects of the extract on human leukemia cells and suggests the role of reactive oxygen species in its cytotoxic activity. Of note, *Withania* induces the expression of specific molecules such as calreticulin, Hsp-70, and Hsp-90 on cancer cells, thus boosting the immunogenic profile of tumor cells and stimulating the innate immune system response. In order to elaborate a preliminary risk/benefit profile, the authors analyze the genotoxicity of the extract through the quantification of histone H2A.X phosphorylation (γ-H2A.X), a biomarker of double-strand DNA breaks. The extract is found to be genotoxic. Bearing in mind that DNA damage plays a well-established role in cancer initiation and poses serious risks for human safety [8], the genotoxicity of *Withania* should be carefully examined for an accurate prediction of its risk–benefit profile.

Burgos-Morón et al. [9] examine the genotoxicity of caffeic acid and a commercial lyophilized coffee extract in cells deficient in the critical DNA repair protein Fanconi anemia D2 and demonstrate that this kind of cell is hypersensitive to the DNA damage induced by caffeic acid and coffee compared to non-deficient cells. These results suggest that coffee and caffeic acid may increase the risk of cancer, particularly in people with germline or sporadic mutations in the DNA repair protein Fanconi anemia D2. Taking into account that caffeic acid accumulates in the urinary bladder, the risk of bladder cancer development may be particularly high. The authors also discuss the key role that caffeic acid and other coffee constituents—such as chlorogenic acid and hydroquinone, endowed with an antioxidant activity at low concentrations and a pro-oxidant activity at higher concentrations—may have in cancer development.

An overview of the effects of some phytoestrogens on cancer progression is presented by Lee et al. [10]. Genistein, resveratrol, kaempferol, and 3,3′-diindolylmethane were extensively studied for their anticancer effects and as alternatives for hormone replacement therapy. In particular, they can inhibit the epithelial–mesenchymal transition, which plays a key role in cancer migration, invasion, and metastasis, and modulate the signaling pathways and the expression of epithelial–mesenchymal transition-related markers, such as TGF-β and PI3K/Akt/mTOR/NF-κB. Nevertheless, phytoestrogens like genistein and resveratrol can have a biphasic effect and lead to cancer cell growth at lower concentrations and to inhibition of cancer cell growth at higher concentrations.

Kakehashi et al. [11] explore the estrogenic effects in the mammary gland and uterus and the carcinogenetic activity of a diet containing *Pueraria mirifica* powder in female rats. To this end, they use different experimental strategies: *Pueraria* administered to ovariectomized animals at doses of 0.03%, 0.3%, and 3% in a phytoestrogen-low diet for 2 weeks; a 4 week application to non-operated rats at a dose of 3% after 7,12-dimethylbenz[a]anthracene cancer initiation; postpubertal administration of 0.3% to 5-week-old non-operated animals for 36 weeks following initiation of mammary and endometrial carcinogenesis with 7,12-dimethylbenz[a]anthracene and *N*-ethyl-*N'*-nitro-*N*-nitrosoguanidine, respectively. In the first experimental setting, *Pueraria* increased uterus weight; in the second one, *Pueraria* stimulated cell proliferation in the mammary gland; in the third experimental model, it boosted mammary adenocarcinoma incidence. These data raise very important questions on the safety of long-term exposure to phytoestrogens with regard to effects on the mammary gland and endometrium. Different products containing *Pueraria mirifica* are widely available in the USA and Japan. Despite the data on its positive health effects, including increasing hair growth, improving appetite, and providing relief for ailments like osteoporosis and even cancer, it evokes an estrogen-like effect that should be considered to better understand its risk–benefit profile. More research has to be performed to better define the relationship between the hazardous and chemopreventive effects of phytoestrogens.

I hope that this Special Issue will provide readers a better understanding of the mechanism of action of phytochemicals in modulating the carcinogenetic process. These aspects have advanced particularly far in recent years, and are extremely useful for the definition of efficient preventive or therapeutic strategies against cancer. I would also like to thank all authors contributing to this Special Issue in Toxins for their commitment and time, and our reviewers for their expert input and critical evaluation of the papers.

References

1. Oh, J.; Hlatky, L.; Jeong, Y.S.; Kim, D. Therapeutic effectiveness of anticancer phytochemicals on cancer stem cells. *Toxins* **2016**, *8*, 199. [CrossRef] [PubMed]
2. Lu, K.T.; Wang, B.Y.; Chi, W.Y.; Chang-Chien, J.; Yang, J.J.; Lee, H.T.; Tzeng, Y.M.; Chang, W.W. Ovatodiolide inhibits breast cancer stem/progenitor cells through SMURF2-mediated downregulation of Hsp27. *Toxins* **2016**, *8*, 127. [CrossRef] [PubMed]
3. Ismail, T.; Calcabrini, C.; Diaz, A.R.; Fimognari, C.; Turrini, E.; Catanzaro, E.; Akhtar, S.; Sestili, P. Ellagitannins in cancer chemoprevention and therapy. *Toxins* **2016**, *8*, 151. [CrossRef] [PubMed]
4. Farhan, M.; Khan, H.Y.; Oves, M.; Al-Harrasi, A.; Rehmani, N.; Arif, H.; Hadi, S.M.; Ahmad, A. Cancer therapy by catechins involves redox cycling of copper ions and generation of reactive oxygen species. *Toxins* **2016**, *8*, 37. [CrossRef] [PubMed]
5. Chen, C.Y.; Yen, C.Y.; Wang, H.R.; Yang, H.P.; Tang, J.Y.; Huang, H.W.; Hsu, S.H.; Chang, H.W. Tenuifolide B from Cinnamomum tenuifolium stem selectively inhibits proliferation of oral cancer cells via apoptosis, ROS generation, mitochondrial depolarization, and DNA damage. *Toxins* **2016**, *8*, 319. [CrossRef] [PubMed]
6. Nguyen, T.T.; Parat, M.O.; Hodson, M.P.; Pan, J.; Shaw, P.N.; Hewavitharana, A.K. Chemical characterization and in vitro cytotoxicity on squamous cell carcinoma cells of Carica papaya leaf extracts. *Toxins* **2016**, *8*, 7. [CrossRef] [PubMed]
7. Turrini, E.; Calcabrini, C.; Sestili, P.; Catanzaro, E.; De Gianni, E.; Diaz, A.R.; Hrelia, P.; Tacchini, M.; Guerrini, A.; Canonico, B.; et al. *Withania somnifera* induces cytotoxic and cytostatic effects on human T leukemia cells. *Toxins* **2016**, *8*, 147. [CrossRef] [PubMed]
8. Turrini, E.; Ferruzzi, L.; Fimognari, C. Natural compounds to overcome cancer chemoresistance: Toxicological and clinical issues. *Expert Opin. Drug Metab. Toxicol.* **2014**, *10*, 1677–1690. [CrossRef] [PubMed]

9.   Burgos-Morón, E.; Calderón-Montaño, J.M.; Orta, M.L.; Guillén-Mancina, E.; Mateos, S.; López-Lázaro, M. Cells deficient in the Fanconi anemia protein FANCD2 are hypersensitive to the cytotoxicity and DNA damage induced by coffee and caffeic acid. *Toxins* **2016**, *8*, 211. [CrossRef] [PubMed]

10. Lee, G.A.; Hwang, K.A.; Choi, K.C. Roles of dietary phytoestrogens on the regulation of epithelial-mesenchymal transition in diverse cancer metastasis. *Toxins* **2016**, *8*, 162. [CrossRef] [PubMed]

11. Kakehashi, A.; Yoshida, M.; Tago, Y.; Ishii, N.; Okuno, T.; Gi, M.; Wanibuchi, H. Pueraria mirifica exerts estrogenic effects in the mammary gland and uterus and promotes mammary carcinogenesis in Donryu rats. *Toxins* **2016**, *8*, 275. [CrossRef] [PubMed]

toxins

MDPI

Review

Therapeutic Effectiveness of Anticancer Phytochemicals on Cancer Stem Cells

Jisun Oh [1],*, Lynn Hlatky [2], Yong-Seob Jeong [3] and Dohoon Kim [4],*

1   School of Food Science and Biotechnology (BK21 Plus), Kyungpook National University, Daegu 41566, Korea
2   Center of Cancer Systems Biology, Tufts University School of Medicine, Boston, MA 02135, USA;
     hlatky@cancer-systems-biology.org
3   Department of Food Science and Technology, Chonbuk National University, Jeonju 54896, Korea;
     ysjeong@jbnu.ac.kr
4   Department of Integrative Physiology and Pathobiology, Tufts University School of Medicine, Boston,
     MA 02111, USA
*   Correspondence: j.oh@knu.ac.kr (J.O.); dohoon.kim@tufts.edu (D.K.); Tel.: +82-53-950-5752 (J.O.);
     +1-617-519-3530 (D.K.)

Academic Editor: Carmela Fimognari
Received: 29 March 2016; Accepted: 23 June 2016; Published: 30 June 2016

Abstract: Understanding how to target cancer stem cells (CSCs) may provide helpful insights for the development of therapeutic or preventive strategies against cancers. Dietary phytochemicals with anticancer properties are promising candidates and have selective impact on CSCs. This review summarizes the influence of phytochemicals on heterogeneous cancer cell populations as well as on specific targeting of CSCs.

Keywords: cancer; cancer stem cells; anticancer; phytochemicals; polyphenols

1. Introduction

While cancer cells are heterogeneous in their tumorigenic potential, a small subset of tumor cells—cancer stem cells (CSCs)—have uniquely high potency for initiating tumorigenesis. These CSCs are postulated to proliferate with unlimited potential, exhibit high resistance to therapy, and have the ability to fuel tumor regrowth post-treatment. Considering the potential of CSCs in both the initial development of cancer and in post-treatment regrowth, they have become a critical focus for the development of new therapeutic strategies. Numerous studies have demonstrated that phytochemicals that have antioxidative properties have anticancer effects. This review summarizes the influence of phytochemicals on cancer cell populations, highlighting the importance of those known to selectively target CSCs and discussing their mechanisms of action.

2. Cancer Stem Cells

Cancer cells within a tumor consist of various clonal subpopulations, thereby exhibiting heterogeneity across many properties, such as genetic variations, marker expression, and proliferative and metastatic potential, and sensitivity to drugs [1]. Considering this heterogeneity of cancer cells, two models have been proposed regarding the origin of tumorigenesis: (i) assembly of diverse cancer clones (referred to as the "stochastic model") and (ii) generation of multiple subclones from a single clone (referred to as the "hierarchical model") [2,3]. In the stochastic model, most cancer cells are capable of proliferating extensively and forming new tumors in cooperation with intrinsic and extrinsic factors. According to this model, tumorigenesis occurs randomly from somatic cells undergoing transformation. In the hierarchical model, on the other hand, only a distinct subpopulation of cancer cells, CSCs, has the ability to extensively proliferate and initiate tumor formation and growth. According to this model, tumorigenesis originates from the CSCs which can be enriched based on unique cellular features [4]. This means only the CSCs possess the cellular capacity to replenish the tumor population. It has recently been shown that cancer non-stem and cancer stem cells may plastically interconvert under particular conditions [5]. However, this does not diminish the fact that the achievement of CSC status, either naturally or by cellular plasticity, is necessary and sufficient for tumorigenicity.

Current treatment approaches in cancer are grounded in the need to kill the majority of cancer cells, based on the stochastic model. However, in many instances where such efforts have not been successful in the treatment of solid cancers, it may be time to refocus our thinking around the hierarchy model in trying to explain resistance to anticancer therapeutics and tumor recurrence [6].

2.1. Cancer Stem Cell Hypothesis

The hierarchical model for tumorigenesis maintains that CSCs are the origin of tumor formation, metastasis, and relapse. In the past two decades, conclusive evidence has demonstrated the existence of CSCs. In 1997, Bonnet and Dick reported a subset of cells—leukemic stem cells—that was isolated from the blood of acute myeloid leukemia, originating from normal hematopoietic stem cells and capable of self-renewing and differentiating into leukemic blasts in immunocompromised mice [7]. This study suggested that the hematopoietic stem cells may be susceptible to leukemic transformation and progression, and was presumably responsible for the hierarchical organization of the leukemic clone. Subsequently, CSCs from solid human tumors, such as breast cancer [8,9], prostate cancer [10] and brain tumors [11], were also isolated and identified on the basis of their tumorigenic capability and cell surface antigen expression. With accumulating evidence for the existence of CSCs within a myriad of other solid tumors [12], the CSC hypothesis has been strongly considered to be a fundamental underpinning of cancer biology that should be considered in thinking about the development of effective cancer therapeutic strategies.

2.2. Cellular Properties of Cancer Stem Cells

Like normal tissue stem cells, CSCs are capable of self-renewal and differentiation into cancer progenitors or mature cancer cells. CSCs can repopulate clonally by cell division (symmetric or asymmetric) or uncontrolled proliferation [13]. Thus, it is thought that CSCs may derive either from normal stem cells that undergo genetic or epigenetic alterations, or from cancer cells (not fully differentiated; cancer progenitor cells) that acquire the potential for unrestrained proliferation [3,14,15]. Although the exact cellular origin of CSCs may be a critical issue in cancer research, it is still unresolved.

In terms of CSC phenotypes, CSCs can be recognized by specific antigens that are expressed, or not, on the cell surface (Table 1). The antibodies against these antigens are generally used for phenotypic characterization or prospective isolation of CSCs (reviewed in [16]). Thus, the specific antigens are considered molecular markers for the validation of CSCs. However, one needs to be cautious in defining the cells expressing these markers as CSCs, since none of these markers are exclusively made by CSCs. Furthermore, CSC phenotypes, even from the same tumor, can exhibit different markers owing to the possible presence of multiple CSC pools, technical variations, or the occurrence of epigenetic alterations [6]. Thus, a combination of different molecular markers, together with epigenetic profiling, may refine the identification of the CSC phenotype.

Table 1. Phenotypic markers for cancer stem cell identification in various tissues.

Tumor	CSC Marker	References
Leukemia	$CD34^+/CD38^-$	[7]
Breast	$CD24^-/CD44^+/Lineage^-/ALDH1^+$	[8,9]
Prostate	$CD44^+/CD133^+/Integrin\ \alpha2\beta1^{high}$	[10,17,18]
Brain	$CD133^+$	[11]
Stomach	$CD44^+/CD133^+$	[19–22]
Pancreas	$CD24^+/CD44^+/CD133^+/ESA^+$	[23–25]

Table 1. *Cont.*

Tumor	CSC Marker	References
Colon	CD44$^+$/CD133$^+$/ALDH1$^+$	[26,27]
Ovary	CD133$^+$/ALDH1$^+$	[28,29]
Lung	CD133$^+$	[30–32]
Liver	CD90$^+$	[33–35]

CSC: cancer stem cell; CD24: heat stable antigen; CD34: hematopoietic progenitor cell antigen; CD38: cyclic ADP ribose hydrolase; CD44: hyaluronate receptor; CD90: Thy-1; CD133: prominin-1; ALDH1: aldehyde dehydrogenase 1A1; ESA: epithelial surface antigen.

Furthermore, tumor tissues composed of malignant cells, including CSCs, reside in the perivascular niche, which is the milieu that nourishes cancer cells and consists of vasculature, hematopoietic cells, inflammatory cells, and myofibroblasts. Although the niche is not indispensable for the sustainment of all types of cancer, mutual interactions between CSCs and the microenvironment are known to profoundly influence cellular properties, such as cell fate and secretory profiles, for certain types of cancer cells [36–38].

3. Anticancer Phytochemicals Targeting CSCs

CSCs are believed to be responsible for the initial formation and growth of cancer, as well as the relapse of cancer after treatment, due to the fact that CSCs are more resistant to conventional therapeutic treatment than differentiated cancer cells [39]. Thus, it would have important implications for cancer prevention and further therapy if treatment could specifically target CSCs while avoiding damage to normal stem cells. Based on their unique features [40,41] and dynamics [42] (reviewed in [6]), CSCs can be targeted by several strategies, such as inhibition of self-renewal, induction of differentiation into mature cancer cells, and sensitization to anticancer agents.

Among approved anticancer drugs, approximately 50% are either natural products or their derivatives [43], primarily from plants, microorganisms, and seeds [44]. Numerous plants have been reported to have anticancer effects [45] or to complement conventional therapeutics by targeting various hallmarks of cancer [46]. Plant-derived natural chemicals, termed phytochemicals, that are used for cancer treatment and that target CSCs are addressed in this section (Figure 1).

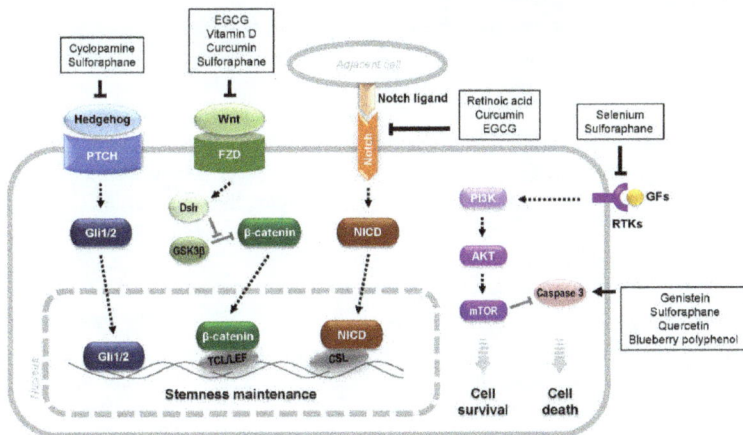

Figure 1. Selected phytochemicals that target signaling pathways involved in stemness maintenance and survival of CSCs. PTCH: Patched, receptors for hedgehog; FZD: Frizzled, receptors for Wnt; Dsh: disheveled, a downstream molecule of FZD; NICD: Notch intracellular domain, a Notch fragment cleaved by γ-secretase; RTKs: receptor tyrosine kinases; GFs: growth factors.

3.1. Anticancer Phytochemicals

A large number of phytochemicals, i.e., chemical compounds produced from plants, including vegetables, fruits, and grains, have been reported to possess anticancer properties and are promoted for cancer prevention and treatment [44,45,47–49] (Table 2). Phytochemicals have been shown to interfere with stabilization of the microtubule structure, thereby inhibiting mitosis and cancer cell propagation. Vincristine and vinblastine, isolated from the leaves of Madagascar periwinkle, were the first phytochemicals to be used clinically in combination with other anticancer agents in lymphomas, leukemias, and breast and lung cancers. Paclitaxel (Taxol), which was originally discovered in the bark of the Pacific yew tree, is one of the most effective and widely used phytochemical compound against breast and ovarian cancers [50,51].

Table 2. Examples of plant-derived anticancer phytochemicals.

Function	Phytochemicals	Plant Derived from
Interference of microtubule stabilization	Vincristine, vinblastine Paclitaxel	Madagascar periwinkle Pacific yew tree
Limitation of cell proliferation	Epigallocatechin-3-gallate Curcumin	*Camellia sinensis* Turmeric
Disruption of chromatin structure	β-lapachone camptothecin podophyllotoxin	Lapacho plant Camptotheca Mayapple plant

Another group of phytochemicals, known as polyphenols, has been shown to have free-radical scavenging activity, working like antioxidants. Epigallocatechin-3-gallate (EGCG), a polyphenol from the leaves of *Camellia sinensis* (processed to green tea), has been used effectively against breast cancer [47]. EGCG was demonstrated to limit cancer cell proliferation by reducing DNA methylation through the inhibition of DNA methyltransferase together with reactivation of the silenced tumor suppressor genes. Curcumin (diferuloylmethane), a polyphenol isolated from the rhizome of the turmeric plant, has also shown therapeutic efficacy on numerous disorders, including cancer [52]. Curcumin is reported to inhibit NF-κB signaling that triggers the intracellular inflammatory response as well as cell-cycle-associated genes. By arresting the cell cycle and inducing apoptosis through the relaying pathways, curcumin interferes with angiogenesis and reduces tumor invasion.

An additional group of anticancer phytochemicals functions as inhibitors of topoisomerase I or II, which are the nuclear enzymes that control DNA supercoiling, eliminate tangles in the chromatin structure, and allow DNA to be replicated and transcribed. Thus, topoisomerase inhibitors can act as anticancer agents by inducing a delay of the cell cycle, followed by cell death [44]. β-Lapachone from the bark of the lapacho plant [53], camptothecin from the bark/stem of *Camptotheca* (the Chinese happy tree), and podophyllotoxin from the root of the Mayapple plant are examples of phytochemicals inhibiting topoisomerases in cancer cells [54,55].

3.2. Phytochemicals Targeting CSCs

Several phytochemicals have been reported to intervene in signaling pathways critical for stemness maintenance of CSCs or to modulate the CSC phenotype [56,57]. The hedgehog, Wnt/β-catenin, and Notch-mediated signaling pathways play important roles in CSC self-renewal and differentiation [58]. Considering that tumorigenesis might be derived from CSCs in which these pathways are aberrantly regulated, the signaling molecules in these pathways may be of particular interest for targeting CSCs [59]. Multiple studies have demonstrated that cancer cell growth can be suppressed by specific inhibitors of these pathways [60,61]. Specific phytochemicals have been reported to influence these signaling pathways. Cyclopamine, initially found in the corn lily (*Veratrum californicum*), targets hedgehog signaling [62–65]. EGCG inhibits Wnt/β-catenin signaling, which affects the self-renewal and invasive abilities of certain CSCs [66–68]. In addition, retinoic acid, the active molecule derived from vitamin A in animals, has been demonstrated to differentiate CSCs or deplete their formation in glioblastoma by downregulating Notch signaling [69,70]. Vitamin D or its analogs can inhibit Notch and/or Wnt/β-catenin signaling and thus induce CSC differentiation [71,72]. Furthermore, curcumin from turmeric and piperine from black and long peppers, well-known anticancer phytochemicals, have also been shown to target breast CSCs by inhibiting Notch and/or Wnt/β-catenin signaling [73]. However, it should be recognized that the inhibition of these self-renewal pathways can affect normal stem cell function as well.

Akt/mTOR signaling is known to be critical for CSC survival and invasion. Akt inhibition causes a preferential induction of apoptosis and reduction of CSC motility [57]. Selenium, as an anticarcinogenic nutrient [74], functions biologically in a form of selenoproteins that are oxidoreductase scavenging oxidants [75]. It was shown that selenium involvement in the modulation of arachidonic acid metabolism could trigger apoptosis of leukemia CSCs [76], and that the apoptosis was regulated through Akt/mTOR signaling [77,78]. However, another study indicated that the biological benefits of selenium supplementation may not necessarily be due to its activity of lowering the level of reactive species [79]. Thus, the exact mechanisms underlying selenium-mediated CSC apoptosis awaits further elucidation. In addition, sulforaphane from cruciferous vegetables, such as broccoli, has been shown to reduce breast and pancreatic CSC viability by affecting Wnt/β-catenin signaling [80,81] or hedgehog signaling [82,83]. Several studies have also demonstrated that sulforaphane can downregulate Akt signaling in various solid cancers [84,85] and breast CSCs [86].

As described above, the polyphenols EGCG and curcumin are known to exert their anticancer effects though antioxidative activity. Polyphenols can inhibit proliferation and/or induce caspase-3-dependent apoptosis of cancer cells via the above-mentioned vital signaling pathways or their cross-talk [87,88]. Being ubiquitously present in nature, polyphenols can be richly extracted from a broad range of fruits, grains, and vegetables, and include flavonoids (categorized into flavones, isoflavones, catechins, and anthocyanins) and lignans. Several studies have suggested that phenolic compounds or polyphenol-containing extracts can influence CSCs as well as cancer cells [89]. Montales et al. reported that the soy isoflavone genistein or blueberry polyphenol treatment could reduce the population of breast CSC-like cells in vitro [90]. Appari et al. showed that a mixture of green tea catechins in combination with sulforaphane and quercetin remarkably inhibited the viability and migration and induced apoptosis of pancreatic CSCs [91]. Lu et al. showed that anthocyanins (i.e., phenolic compounds found in grapes, eggplants, red cabbages, and radishes) can inhibit cancer invasion and epithelial-mesenchymal transition of uterine cervical cancer cells [92]. Quercetin, a flavonol that can be enriched from apples, onions, teas, and berries, has demonstrated efficacy against pancreatic and head/neck CSCs [93,94]. The synergistic effect of quercetin with EGCG or sulforaphane in eliminating prostate or pancreatic CSCs has also been described [95,96]. Thus, polyphenols, including flavonoids, may be considered promising anticancer agents for targeting various CSCs, although further extensive investigation on their bioavailability and working mechanisms at the cellular and molecular levels have yet to be done.

On the basis of the accumulating evidence supporting the anticancer activity of phytochemicals, there have been some clinical studies. Recent clinical trials have shown the effectiveness of curcumin, green tea catechins, including EGCG, and sulforaphane against various cancers (reviewed in [97,98]). In particular, curcumin showed its therapeutic potency in human clinical trials [99] via targeting CSCs [100]. However, the low bioavailability of curcumin makes its therapeutic use challenging. To overcome this issue, several strategies such as structural modifications or special formulations are being tried. It is expected that in vitro and in vivo functional studies using the compounds or their derivatives targeting CSCs may provide useful information on eliminating CSCs and inhibiting tumorigenesis.

Along with the emerging evidence supporting the beneficial effects against CSCs as well as cancer cells, the beauty of phytochemicals is the fact that they are naturally present in edible plant materials. This warrants the assurance of safety for ingestion. In addition, certain phytochemicals sensitize CSCs to conventional chemotherapeutic agents by interfering with the key signaling pathways for cell survival, stemness maintenance of CSCs, or both. Thus, synergistic effects are expected when the CSC-targeting phytochemicals and chemotherapeutic drugs are used combinatorially [98]. However, rigorous examinations are required to test the potential adverse effects, such as the counteraction of phytochemicals against chemotherapeutic drugs and the additive toxicity of phytochemicals [101].

4. Summary and Conclusions

Tumors comprise of phenotypically and functionally heterogeneous cells. Based on this feature, two models have been established regarding tumorigenesis: the stochastic model and the hierarchical model. The latter model postulates a hierarchical organization of diverse populations of cells and premises the presence of CSCs that account for the sustainment of tumorigenesis. Thus, control of CSCs may be a necessary first step in an effective strategy for cancer treatment. Dietary phytochemicals can exert influence at all stages of cancer development. Since some of the phytochemicals are already known to affect CSC viability and fate, extensive and intensive studies of these compounds should provide insight into their pharmaceutical efficacy for cancer prevention and therapy.

Acknowledgments: This study was supported by the Basic Science Research Program through the National Research Foundation of Korea funded by the Ministry of Education (Grant No. NRF-2013R1A1A2013362).

Conflicts of Interest: The authors declare no conflict of interest.

References

1. Heppner, G.H.; Miller, B.E. Tumor heterogeneity: Biological implications and therapeutic consequences. *Cancer Metastasis Rev.* **1983**, *2*, 5–23. [CrossRef] [PubMed]
2. Dick, J.E. Looking ahead in cancer stem cell research. *Nat. Biotechnol.* **2009**, *27*, 44–46. [CrossRef] [PubMed]
3. Reya, T.; Morrison, S.J.; Clarke, M.F.; Weissman, I.L. Stem cells, cancer, and cancer stem cells. *Nature* **2001**, *414*, 105–111. [CrossRef] [PubMed]
4. Dick, J.E. Stem cell concepts renew cancer research. *Blood* **2008**, *112*, 4793–4807. [CrossRef] [PubMed]
5. Cabrera, M.C.; Hollingsworth, R.E.; Hurt, E.M. Cancer stem cell plasticity and tumor hierarchy. *World J. Stem Cells* **2015**, *7*, 27–36. [CrossRef] [PubMed]
6. Visvader, J.E.; Lindeman, G.J. Cancer stem cells: Current status and evolving complexities. *Cell Stem Cell* **2012**, *10*, 717–728. [CrossRef] [PubMed]
7. Bonnet, D.; Dick, J.E. Human acute myeloid leukemia is organized as a hierarchy that originates from a primitive hematopoietic cell. *Nat. Med.* **1997**, *3*, 730–737. [CrossRef] [PubMed]
8. Al-Hajj, M.; Wicha, M.S.; Benito-Hernandez, A.; Morrison, S.J.; Clarke, M.F. Prospective identification of tumorigenic breast cancer cells. *Proc. Natl. Acad. Sci. USA* **2003**, *100*, 3983–3988. [CrossRef] [PubMed]
9. Ginestier, C.; Hur, M.H.; Charafe-Jauffret, E.; Monville, F.; Dutcher, J.; Brown, M.; Jacquemier, J.; Viens, P.; Kleer, C.G.; Liu, S.; et al. ALDH1 is a marker of normal and malignant human mammary stem cells and a predictor of poor clinical outcome. *Cell Stem Cell* **2007**, *1*, 555–567. [CrossRef] [PubMed]
10. Collins, A.T.; Berry, P.A.; Hyde, C.; Stower, M.J.; Maitland, N.J. Prospective identification of tumorigenic prostate cancer stem cells. *Cancer Res.* **2005**, *65*, 10946–10951. [CrossRef] [PubMed]
11. Singh, S.K.; Hawkins, C.; Clarke, I.D.; Squire, J.A.; Bayani, J.; Hide, T.; Henkelman, R.M.; Cusimano, M.D.; Dirks, P.B. Identification of human brain tumour initiating cells. *Nature* **2004**, *432*, 396–401. [CrossRef] [PubMed]
12. Visvader, J.E.; Lindeman, G.J. Cancer stem cells in solid tumours: Accumulating evidence and unresolved questions. *Nat. Rev. Cancer* **2008**, *8*, 755–768. [CrossRef] [PubMed]
13. Li, Y.; Wicha, M.S.; Schwartz, S.J.; Sun, D. Implications of cancer stem cell theory for cancer chemoprevention by natural dietary compounds. *J. Nutr. Biochem.* **2011**, *22*, 799–806. [CrossRef] [PubMed]
14. Takebe, N.; Ivy, S.P. Controversies in cancer stem cells: Targeting embryonic signaling pathways. *Clin. Cancer Res.* **2010**, *16*, 3106–3112. [CrossRef] [PubMed]

15. Eaves, C.J. Cancer stem cells: Here, there, everywhere? *Nature* **2008**, *456*, 581–582. [CrossRef] [PubMed]
16. Kreso, A.; Dick, J.E. Evolution of the cancer stem cell model. *Cell Stem Cell* **2014**, *14*, 275–291. [CrossRef] [PubMed]
17. Vander Griend, D.J.; Karthaus, W.L.; Dalrymple, S.; Meeker, A.; DeMarzo, A.M.; Isaacs, J.T. The role of CD133 in normal human prostate stem cells and malignant cancer-initiating cells. *Cancer Res.* **2008**, *68*, 9703–9711. [CrossRef] [PubMed]
18. Williamson, S.C.; Hepburn, A.C.; Wilson, L.; Coffey, K.; Ryan-Munden, C.A.; Pal, D.; Leung, H.Y.; Robson, C.N.; Heer, R. Human $\alpha_2 \beta_1^{HI}$ CD133^{+VE} epithelial prostate stem cells express low levels of active androgen receptor. *PLoS One* **2012**, *7*, e48944. [CrossRef] [PubMed]
19. Wang, T.; Ong, C.W.; Shi, J.; Srivastava, S.; Yan, B.; Cheng, C.L.; Yong, W.P.; Chan, S.L.; Yeoh, K.G.; Iacopetta, B.; et al. Sequential expression of putative stem cell markers in gastric carcinogenesis. *Br. J. Cancer* **2011**, *105*, 658–665. [CrossRef] [PubMed]
20. Zhang, C.; Li, C.; He, F.; Cai, Y.; Yang, H. Identification of CD44+CD24+ gastric cancer stem cells. *J. Cancer Res. Clin. Oncol.* **2011**, *137*, 1679–1686. [CrossRef] [PubMed]
21. Chen, W.; Zhang, X.; Chu, C.; Cheung, W.L.; Ng, L.; Lam, S.; Chow, A.; Lau, T.; Chen, M.; Li, Y.; et al. Identification of CD44+ cancer stem cells in human gastric cancer. *Hepatogastroenterology* **2013**, *60*, 949–954. [PubMed]
22. Chen, S.; Hou, J.H.; Feng, X.Y.; Zhang, X.S.; Zhou, Z.W.; Yun, J.P.; Chen, Y.B.; Cai, M.Y. Clinicopathologic significance of putative stem cell marker, CD44 and CD133, in human gastric carcinoma. *J. Surg. Oncol.* **2013**, *107*, 799–806. [CrossRef] [PubMed]
23. Hermann, P.C.; Huber, S.L.; Herrler, T.; Aicher, A.; Ellwart, J.W.; Guba, M.; Bruns, C.J.; Heeschen, C. Distinct populations of cancer stem cells determine tumor growth and metastatic activity in human pancreatic cancer. *Cell Stem Cell* **2007**, *1*, 313–323. [CrossRef] [PubMed]
24. Fitzgerald, T.L.; McCubrey, J.A. Pancreatic cancer stem cells: Association with cell surface markers, prognosis, resistance, metastasis and treatment. *Adv. Biol. Regul.* **2014**, *56*, 45–50. [CrossRef] [PubMed]
25. Li, C.; Heidt, D.G.; Dalerba, P.; Burant, C.F.; Zhang, L.; Adsay, V.; Wicha, M.; Clarke, M.F.; Simeone, D.M. Identification of pancreatic cancer stem cells. *Cancer. Res.* **2007**, *67*, 1030–1037. [CrossRef] [PubMed]
26. Ricci-Vitiani, L.; Lombardi, D.G.; Pilozzi, E.; Biffoni, M.; Todaro, M.; Peschle, C.; De Maria, R. Identification and expansion of human colon-cancer-initiating cells. *Nature* **2007**, *445*, 111–115. [CrossRef] [PubMed]
27. O'Brien, C.A.; Pollett, A.; Gallinger, S.; Dick, J.E. A human colon cancer cell capable of initiating tumour growth in immunodeficient mice. *Nature* **2007**, *445*, 106–110. [CrossRef] [PubMed]
28. Silva, I.A.; Bai, S.; McLean, K.; Yang, K.; Griffith, K.; Thomas, D.; Ginestier, C.; Johnston, C.; Kueck, A.; Reynolds, R.K.; et al. Aldehyde dehydrogenase in combination with CD133 defines angiogenic ovarian cancer stem cells that portend poor patient survival. *Cancer Res.* **2011**, *71*, 3991–4001. [CrossRef] [PubMed]
29. Kryczek, I.; Liu, S.; Roh, M.; Vatan, L.; Szeliga, W.; Wei, S.; Banerjee, M.; Mao, Y.; Kotarski, J.; Wicha, M.S.; et al. Expression of aldehyde dehydrogenase and CD133 defines ovarian cancer stem cells. *Int. J. Cancer* **2012**, *130*, 29–39. [CrossRef] [PubMed]
30. Eramo, A.; Lotti, F.; Sette, G.; Pilozzi, E.; Biffoni, M.; Di Virgilio, A.; Conticello, C.; Ruco, L.; Peschle, C.; De Maria, R. Identification and expansion of the tumorigenic lung cancer stem cell population. *Cell Death Differ.* **2008**, *15*, 504–514. [CrossRef] [PubMed]
31. Bertolini, G.; Roz, L.; Perego, P.; Tortoreto, M.; Fontanella, E.; Gatti, L.; Pratesi, G.; Fabbri, A.; Andriani, F.; Tinelli, S.; et al. Highly tumorigenic lung cancer CD133+ cells display stem-like features and are spared by cisplatin treatment. *Proc. Natl. Acad. Sci. USA* **2009**, *106*, 16281–16286. [CrossRef] [PubMed]
32. Tirino, V.; Camerlingo, R.; Franco, R.; Malanga, D.; La Rocca, A.; Viglietto, G.; Rocco, G.; Pirozzi, G. The role of CD133 in the identification and characterisation of tumour-initiating cells in non-small-cell lung cancer. *Eur. J. Cardiothorac. Surg.* **2009**, *36*, 446–453. [CrossRef] [PubMed]
33. Yang, Z.F.; Ho, D.W.; Ng, M.N.; Lau, C.K.; Yu, W.C.; Ngai, P.; Chu, P.W.; Lam, C.T.; Poon, R.T.; Fan, S.T. Significance of CD90+ cancer stem cells in human liver cancer. *Cancer Cell* **2008**, *13*, 153–166. [CrossRef] [PubMed]
34. Yang, Z.F.; Ngai, P.; Ho, D.W.; Yu, W.C.; Ng, M.N.; Lau, C.K.; Li, M.L.; Tam, K.H.; Lam, C.T.; Poon, R.T.; et al. Identification of local and circulating cancer stem cells in human liver cancer. *Hepatology* **2008**, *47*, 919–928. [CrossRef] [PubMed]

35. Tomuleasa, C.; Soritau, O.; Rus-Ciuca, D.; Pop, T.; Todea, D.; Mosteanu, O.; Pintea, B.; Foris, V.; Susman, S.; Kacso, G.; et al. Isolation and characterization of hepatic cancer cells with stem-like properties from hepatocellular carcinoma. *J. Gastrointestin. Liver Dis.* **2010**, *19*, 61–67. [PubMed]

36. Gilbertson, R.J.; Rich, J.N. Making a tumour's bed: Glioblastoma stem cells and the vascular niche. *Nat. Rev. Cancer* **2007**, *7*, 733–736. [CrossRef] [PubMed]

37. Kelly, P.N.; Dakic, A.; Adams, J.M.; Nutt, S.L.; Strasser, A. Tumor growth need not be driven by rare cancer stem cells. *Science* **2007**, *317*, 337. [CrossRef] [PubMed]

38. Vermeulen, L.; De Sousa, E.M.F.; van der Heijden, M.; Cameron, K.; de Jong, J.H.; Borovski, T.; Tuynman, J.B.; Todaro, M.; Merz, C.; Rodermond, H.; et al. Wnt activity defines colon cancer stem cells and is regulated by the microenvironment. *Nat. Cell. Biol.* **2010**, *12*, 468–476. [CrossRef] [PubMed]

39. Creighton, C.J.; Li, X.; Landis, M.; Dixon, J.M.; Neumeister, V.M.; Sjolund, A.; Rimm, D.L.; Wong, H.; Rodriguez, A.; Herschkowitz, J.I.; et al. Residual breast cancers after conventional therapy display mesenchymal as well as tumor-initiating features. *Proc. Natl. Acad. Sci. USA* **2009**, *106*, 13820–13825. [CrossRef] [PubMed]

40. Saito, Y.; Uchida, N.; Tanaka, S.; Suzuki, N.; Tomizawa-Murasawa, M.; Sone, A.; Najima, Y.; Takagi, S.; Aoki, Y.; Wake, A.; et al. Induction of cell cycle entry eliminates human leukemia stem cells in a mouse model of AML. *Nat. Biotechnol.* **2010**, *28*, 275–280. [PubMed]

41. Lathia, J.D.; Hitomi, M.; Gallagher, J.; Gadani, S.P.; Adkins, J.; Vasanji, A.; Liu, L.; Eyler, C.E.; Heddleston, J.M.; Wu, Q.; et al. Distribution of CD133 reveals glioma stem cells self-renew through symmetric and asymmetric cell divisions. *Cell Death Dis.* **2011**, *2*, e200. [CrossRef] [PubMed]

42. Fornari, C.; Beccuti, M.; Lanzardo, S.; Conti, L.; Balbo, G.; Cavallo, F.; Calogero, R.A.; Cordero, F. A mathematical-biological joint effort to investigate the tumor-initiating ability of Cancer Stem Cells. *PLoS ONE* **2014**, *9*, e106193. [CrossRef] [PubMed]

43. Newman, D.J.; Cragg, G.M. Natural products as sources of new drugs over the 30 years from 1981 to 2010. *J. Nat. Prod.* **2012**, *75*, 311–335. [CrossRef] [PubMed]

44. Nobili, S.; Lippi, D.; Witort, E.; Donnini, M.; Bausi, L.; Mini, E.; Capaccioli, S. Natural compounds for cancer treatment and prevention. *Pharmacol. Res.* **2009**, *59*, 365–378. [CrossRef] [PubMed]

45. Graham, J.G.; Quinn, M.L.; Fabricant, D.S.; Farnsworth, N.R. Plants used against cancer - an extension of the work of Jonathan Hartwell. *J. Ethnopharmacol.* **2000**, *73*, 347–377. [CrossRef]

46. Bishayee, A.; Block, K. A broad-spectrum integrative design for cancer prevention and therapy: The challenge ahead. *Semin. Cancer Biol.* **2015**, *35*, S1–S4. [CrossRef] [PubMed]

47. Amin, A.; Gali-Muhtasib, H.; Ocker, M.; Schneider-Stock, R. Overview of major classes of plant-derived anticancer drugs. *Int. J. Biomed. Sci.* **2009**, *5*, 1–11. [PubMed]

48. Diederich, M.; Cerella, C. Non-canonical programmed cell death mechanisms triggered by natural compounds. *Semin. Cancer Biol.* **2016**. [CrossRef] [PubMed]

49. Shanmugam, M.K.; Lee, J.H.; Chai, E.Z.; Kanchi, M.M.; Kar, S.; Arfuso, F.; Dharmarajan, A.; Kumar, A.P.; Ramar, P.S.; Looi, C.Y.; et al. Cancer prevention and therapy through the modulation of transcription factors by bioactive natural compounds. *Semin. Cancer Biol.* **2016**. [CrossRef] [PubMed]

50. Wani, M.C.; Taylor, H.L.; Wall, M.E.; Coggon, P.; McPhail, A.T. Plant antitumor agents. VI. The isolation and structure of taxol, a novel antileukemic and antitumor agent from Taxus brevifolia. *J. Am. Chem. Soc.* **1971**, *93*, 2325–2327. [CrossRef] [PubMed]

51. Schiff, P.B.; Fant, J.; Horwitz, S.B. Promotion of microtubule assembly in vitro by taxol. *Nature* **1979**, *277*, 665–667. [CrossRef] [PubMed]

52. Sa, G.; Das, T. Anti cancer effects of curcumin: Cycle of life and death. *Cell Div.* **2008**, *3*. [CrossRef] [PubMed]

53. Li, Y.Z.; Li, C.J.; Pinto, A.V.; Pardee, A.B. Release of mitochondrial cytochrome C in both apoptosis and necrosis induced by beta-lapachone in human carcinoma cells. *Mol. Med.* **1999**, *5*, 232–239. [PubMed]

54. Pommier, Y. Topoisomerase I inhibitors: Camptothecins and beyond. *Nat. Rev. Cancer* **2006**, *6*, 789–802. [CrossRef] [PubMed]

55. Hartmann, J.T.; Lipp, H.P. Camptothecin and podophyllotoxin derivatives: Inhibitors of topoisomerase I and II - mechanisms of action, pharmacokinetics and toxicity profile. *Drug Saf.* **2006**, *29*, 209–230. [CrossRef] [PubMed]

56. Kim, D.H.; Surh, Y.J. Chemopreventive and therapeutic ootential of phytochemicals targeting cancer stem cells. *Curr. Pharmacol. Rep.* **2015**, *1*, 302–311. [CrossRef]

57. Dandawate, P.; Padhye, S.; Ahmad, A.; Sarkar, F.H. Novel strategies targeting cancer stem cells through phytochemicals and their analogs. *Drug Deliv. Transl. Res.* **2013**, *3*, 165–182. [CrossRef] [PubMed]

58. Kim, Y.S.; Farrar, W.; Colburn, N.H.; Milner, J.A. Cancer stem cells: Potential target for bioactive food components. *J. Nutr. Biochem.* **2012**, *23*, 691–698. [CrossRef] [PubMed]

59. Liu, S.; Dontu, G.; Wicha, M.S. Mammary stem cells, self-renewal pathways, and carcinogenesis. *Breast. Cancer Res.* **2005**, *7*, 86–95. [CrossRef] [PubMed]

60. Kubo, M.; Nakamura, M.; Tasaki, A.; Yamanaka, N.; Nakashima, H.; Nomura, M.; Kuroki, S.; Katano, M. Hedgehog signaling pathway is a new therapeutic target for patients with breast cancer. *Cancer Res.* **2004**, *64*, 6071–6074. [CrossRef] [PubMed]

61. Romer, J.T.; Kimura, H.; Magdaleno, S.; Sasai, K.; Fuller, C.; Baines, H.; Connelly, M.; Stewart, C.F.; Gould, S.; Rubin, L.L.; et al. Suppression of the Shh pathway using a small molecule inhibitor eliminates medulloblastoma in *Ptc1$^{+/-}$p53$^{-/-}$* mice. *Cancer Cell* **2004**, *6*, 229–240. [CrossRef] [PubMed]

62. Berman, D.M.; Karhadkar, S.S.; Hallahan, A.R.; Pritchard, J.I.; Eberhart, C.G.; Watkins, D.N.; Chen, J.K.; Cooper, M.K.; Taipale, J.; Olson, J.M.; et al. Medulloblastoma growth inhibition by hedgehog pathway blockade. *Science* **2002**, *297*, 1559–1561. [CrossRef] [PubMed]

63. Liu, S.; Dontu, G.; Mantle, I.D.; Patel, S.; Ahn, N.S.; Jackson, K.W.; Suri, P.; Wicha, M.S. Hedgehog signaling and Bmi-1 regulate self-renewal of normal and malignant human mammary stem cells. *Cancer Res.* **2006**, *66*, 6063–6071. [CrossRef] [PubMed]

64. Feldmann, G.; Dhara, S.; Fendrich, V.; Bedja, D.; Beaty, R.; Mullendore, M.; Karikari, C.; Alvarez, H.; Iacobuzio-Donahue, C.; Jimeno, A.; et al. Blockade of hedgehog signaling inhibits pancreatic cancer invasion and metastases: A new paradigm for combination therapy in solid cancers. *Cancer Res.* **2007**, *67*, 2187–2196. [CrossRef] [PubMed]

65. Peacock, C.D.; Wang, Q.; Gesell, G.S.; Corcoran-Schwartz, I.M.; Jones, E.; Kim, J.; Devereux, W.L.; Rhodes, J.T.; Huff, C.A.; Beachy, P.A.; et al. Hedgehog signaling maintains a tumor stem cell compartment in multiple myeloma. *Proc. Natl. Acad. Sci. USA* **2007**, *104*, 4048–4053. [CrossRef] [PubMed]

66. Lee, S.H.; Nam, H.J.; Kang, H.J.; Kwon, H.W.; Lim, Y.C. Epigallocatechin-3-gallate attenuates head and neck cancer stem cell traits through suppression of Notch pathway. *Eur. J. Cancer* **2013**, *49*, 3210–3218. [CrossRef] [PubMed]

67. Mineva, N.D.; Paulson, K.E.; Naber, S.P.; Yee, A.S.; Sonenshein, G.E. Epigallocatechin-3-gallate inhibits stem-like inflammatory breast cancer cells. *PLoS ONE* **2013**, *8*, e73464. [CrossRef] [PubMed]

68. Lin, C.H.; Shen, Y.A.; Hung, P.H.; Yu, Y.B.; Chen, Y.J. Epigallocathechin gallate, polyphenol present in green tea, inhibits stem-like characteristics and epithelial-mesenchymal transition in nasopharyngeal cancer cell lines. *BMC Complement. Altern. Med.* **2012**, *12*. [CrossRef] [PubMed]

69. Clarke, N.; Germain, P.; Altucci, L.; Gronemeyer, H. Retinoids: Potential in cancer prevention and therapy. *Expert Rev. Mol. Med.* **2004**, *6*, 1–23. [CrossRef] [PubMed]

70. Ying, M.; Wang, S.; Sang, Y.; Sun, P.; Lal, B.; Goodwin, C.R.; Guerrero-Cazares, H.; Quinones-Hinojosa, A.; Laterra, J.; Xia, S. Regulation of glioblastoma stem cells by retinoic acid: Role for Notch pathway inhibition. *Oncogene* **2011**, *30*, 3454–3467. [CrossRef] [PubMed]

71. Palmer, H.G.; Gonzalez-Sancho, J.M.; Espada, J.; Berciano, M.T.; Puig, I.; Baulida, J.; Quintanilla, M.; Cano, A.; de Herreros, A.G.; Lafarga, M.; et al. Vitamin D$_3$ promotes the differentiation of colon carcinoma cells by the induction of E-cadherin and the inhibition of beta-catenin signaling. *J. Cell Biol.* **2001**, *154*, 369–387. [CrossRef] [PubMed]

72. Garcia, J.J.; Lopez-Pingarron, L.; Almeida-Souza, P.; Tres, A.; Escudero, P.; Garcia-Gil, F.A.; Tan, D.X.; Reiter, R.J.; Ramirez, J.M.; Bernal-Perez, M. Protective effects of melatonin in reducing oxidative stress and in preserving the fluidity of biological membranes: A review. *J. Pineal Res.* **2014**, *56*, 225–237. [CrossRef] [PubMed]

73. Kakarala, M.; Brenner, D.E.; Korkaya, H.; Cheng, C.; Tazi, K.; Ginestier, C.; Liu, S.; Dontu, G.; Wicha, M.S. Targeting breast stem cells with the cancer preventive compounds curcumin and piperine. *Breast Cancer Res. Treat.* **2010**, *122*, 777–785. [CrossRef] [PubMed]

74. Zeng, H.; Combs, G.F., Jr. Selenium as an anticancer nutrient: Roles in cell proliferation and tumor cell invasion. *J. Nutr. Biochem.* **2008**, *19*, 1–7. [CrossRef] [PubMed]

75. Hatfield, D.L.; Tsuji, P.A.; Carlson, B.A.; Gladyshev, V.N. Selenium and selenocysteine: Roles in cancer, health, and development. *Trends Biochem. Sci.* **2014**, *39*, 112–120. [CrossRef] [PubMed]

76. Gandhi, U.H.; Kaushal, N.; Hegde, S.; Finch, E.R.; Kudva, A.K.; Kennett, M.J.; Jordan, C.T.; Paulson, R.F.; Prabhu, K.S. Selenium suppresses leukemia through the action of endogenous eicosanoids. *Cancer Res.* **2014**, *74*, 3890–3901. [CrossRef] [PubMed]

77. Sanmartin, C.; Plano, D.; Sharma, A.K.; Palop, J.A. Selenium compounds, apoptosis and other types of cell death: An overview for cancer therapy. *Int. J. Mol. Sci.* **2012**, *13*, 9649–9672. [CrossRef] [PubMed]

78. Kim, I.; He, Y.Y. Targeting the AMP-Activated Protein Kinase for Cancer Prevention and Therapy. *Front. Oncol.* **2013**, *3*. [CrossRef] [PubMed]

79. Li, F.; Lutz, P.B.; Pepelyayeva, Y.; Arner, E.S.; Bayse, C.A.; Rozovsky, S. Redox active motifs in selenoproteins. *Proc. Natl. Acad. Sci. USA* **2014**, *111*, 6976–6981. [CrossRef] [PubMed]

80. Li, Y.; Zhang, T.; Korkaya, H.; Liu, S.; Lee, H.F.; Newman, B.; Yu, Y.; Clouthier, S.G.; Schwartz, S.J.; Wicha, M.S.; et al. Sulforaphane, a dietary component of broccoli/broccoli sprouts, inhibits breast cancer stem cells. *Clin. Cancer Res.* **2010**, *16*, 2580–2590. [CrossRef] [PubMed]

81. Kallifatidis, G.; Rausch, V.; Baumann, B.; Apel, A.; Beckermann, B.M.; Groth, A.; Mattern, J.; Li, Z.; Kolb, A.; Moldenhauer, G.; et al. Sulforaphane targets pancreatic tumour-initiating cells by NF-kappaB-induced antiapoptotic signalling. *Gut* **2009**, *58*, 949–963. [CrossRef] [PubMed]

82. Filipe, P.; Morliere, P.; Silva, J.N.; Maziere, J.C.; Patterson, L.K.; Freitas, J.P.; Santus, R. Plasma lipoproteins as mediators of the oxidative stress induced by UV light in human skin: A review of biochemical and biophysical studies on mechanisms of apolipoprotein alteration, lipid peroxidation, and associated skin cell responses. *Oxidative Med. Cell. Longevity* **2013**, *2013*, 285825. [CrossRef] [PubMed]

83. Rodova, M.; Fu, J.; Watkins, D.N.; Srivastava, R.K.; Shankar, S. Sonic hedgehog signaling inhibition provides opportunities for targeted therapy by sulforaphane in regulating pancreatic cancer stem cell self-renewal. *PLoS ONE* **2012**, *7*, e46083. [CrossRef] [PubMed]

84. Chaudhuri, D.; Orsulic, S.; Ashok, B.T. Antiproliferative activity of sulforaphane in Akt-overexpressing ovarian cancer cells. *Mol. Cancer Ther.* **2007**, *6*, 334–345. [CrossRef] [PubMed]

85. Shankar, S.; Ganapathy, S.; Srivastava, R.K. Sulforaphane enhances the therapeutic potential of TRAIL in prostate cancer orthotopic model through regulation of apoptosis, metastasis, and angiogenesis. *Clin. Cancer Res.* **2008**, *14*, 6855–6866. [CrossRef] [PubMed]

86. Korkaya, H.; Paulson, A.; Charafe-Jauffret, E.; Ginestier, C.; Brown, M.; Dutcher, J.; Clouthier, S.G.; Wicha, M.S. Regulation of mammary stem/progenitor cells by PTEN/Akt/beta-catenin signaling. *PLoS Biol.* **2009**, *7*, e1000121. [CrossRef] [PubMed]

87. Fresco, P.; Borges, F.; Diniz, C.; Marques, M.P. New insights on the anticancer properties of dietary polyphenols. *Med. Res. Rev.* **2006**, *26*, 747–766. [CrossRef] [PubMed]

88. Ramos, S. Cancer chemoprevention and chemotherapy: Dietary polyphenols and signalling pathways. *Mol. Nutr. Food Res.* **2008**, *52*, 507–526. [CrossRef] [PubMed]

89. Sak, K.; Everaus, H. Role of Flavonoids in Future Anticancer Therapy by Eliminating the Cancer Stem Cells. *Curr. Stem Cell Res. Ther.* **2015**, *10*, 271–282. [CrossRef] [PubMed]

90. Montales, M.T.; Rahal, O.M.; Kang, J.; Rogers, T.J.; Prior, R.L.; Wu, X.; Simmen, R.C. Repression of mammosphere formation of human breast cancer cells by soy isoflavone genistein and blueberry polyphenolic acids suggests diet-mediated targeting of cancer stem-like/progenitor cells. *Carcinogenesis* **2012**, *33*, 652–660. [CrossRef] [PubMed]

91. Appari, M.; Babu, K.R.; Kaczorowski, A.; Gross, W.; Herr, I. Sulforaphane, quercetin and catechins complement each other in elimination of advanced pancreatic cancer by miR-let-7 induction and K-ras inhibition. *Int. J. Oncol.* **2014**, *45*, 1391–1400. [CrossRef] [PubMed]

92. Lu, J.N.; Lee, W.S.; Yun, J.W.; Kim, M.J.; Kim, H.J.; Kim, D.C.; Jeong, J.H.; Choi, Y.H.; Kim, G.S.; Ryu, C.H.; et al. Anthocyanins from Vitis coignetiae Pulliat Inhibit Cancer Invasion and Epithelial-Mesenchymal Transition, but These Effects Can Be Attenuated by Tumor Necrosis Factor in Human Uterine Cervical Cancer HeLa Cells. *Evid. Based Complement. Altern. Med.* **2013**, *2013*, 503043. [CrossRef] [PubMed]

93. Chang, W.W.; Hu, F.W.; Yu, C.C.; Wang, H.H.; Feng, H.P.; Lan, C.; Tsai, L.L.; Chang, Y.C. Quercetin in elimination of tumor initiating stem-like and mesenchymal transformation property in head and neck cancer. *Head Neck* **2013**, *35*, 413–419. [CrossRef] [PubMed]

94. Zhou, W.; Kallifatidis, G.; Baumann, B.; Rausch, V.; Mattern, J.; Gladkich, J.; Giese, N.; Moldenhauer, G.; Wirth, T.; Buchler, M.W.; et al. Dietary polyphenol quercetin targets pancreatic cancer stem cells. *Int. J. Oncol.* **2010**, *37*, 551–561. [PubMed]

95. Srivastava, R.K.; Tang, S.N.; Zhu, W.; Meeker, D.; Shankar, S. Sulforaphane synergizes with quercetin to inhibit self-renewal capacity of pancreatic cancer stem cells. *Front. Biosci.* **2011**, *3*, 515–528. [CrossRef]

96. Tang, S.N.; Singh, C.; Nall, D.; Meeker, D.; Shankar, S.; Srivastava, R.K. The dietary bioflavonoid quercetin synergizes with epigallocathechin gallate (EGCG) to inhibit prostate cancer stem cell characteristics, invasion, migration and epithelial-mesenchymal transition. *J. Mol. Signal.* **2010**, *5*, 14. [CrossRef] [PubMed]

97. Hosseini, A.; Ghorbani, A. Cancer therapy with phytochemicals: Evidence from clinical studies. *Avicenna J. Phytomed.* **2015**, *5*, 84–97. [PubMed]

98. Scarpa, E.S.; Ninfali, P. Phytochemicals as Innovative Therapeutic Tools against Cancer Stem Cells. *Int. J. Mol. Sci.* **2015**, *16*, 15727–15742. [CrossRef] [PubMed]

99. Hatcher, H.; Planalp, R.; Cho, J.; Torti, F.M.; Torti, S.V. Curcumin: From ancient medicine to current clinical trials. *Cell. Mol. Life Sci.* **2008**, *65*, 1631–1652. [CrossRef] [PubMed]

100. Li, Y.; Zhang, T. Targeting cancer stem cells by curcumin and clinical applications. *Cancer Lett.* **2014**, *346*, 197–205. [CrossRef] [PubMed]

101. Sparreboom, A.; Cox, M.C.; Acharya, M.R.; Figg, W.D. Herbal remedies in the United States: Potential adverse interactions with anticancer agents. *J. Clin. Oncol.* **2004**, *22*, 2489–2503. [CrossRef] [PubMed]

toxins

MDPI

Article

Ovatodiolide Inhibits Breast Cancer Stem/Progenitor Cells through SMURF2-Mediated Downregulation of Hsp27

Kuan-Ta Lu [1], Bing-Yen Wang [2,3,4], Wan-Yu Chi [5], Ju Chang-Chien [5,6], Jiann-Jou Yang [5,7], Hsueh-Te Lee [8], Yew-Min Tzeng [9,10,*] and Wen-Wei Chang [5,7,*]

[1]   Department of Anesthesiology, Changhua Christian Hospital, Changhua 500, Taiwan; 97343@cch.org.tw
[2]   Division of Thoracic Surgery, Department of Surgery, Changhua Christian Hospital, Changhua 500, Taiwan; 156283@cch.org.tw
[3]   School of Medicine, Kaohsiung Medical University, Kaohsiung 807, Taiwan
[4]   Institute of Genomics and Bioinformatics, National Chung Hsing University, Taichung 402, Taiwan
[5]   School of Biomedical Sciences, Chung Shan Medical University, Taichung 40201, Taiwan; zx82704@yahoo.com.tw (W.-Y.C.); swan1204@hotmail.com (J.C.-C.); jiannjou@csmu.edu.tw (J.-J.Y.)
[6]   Institute of Microbiology & Immunology, Chung Shan Medical University, Taichung 40201, Taiwan
[7]   Department of Medical Research, Chung Shan Medical University Hospital, Taichung 40201, Taiwan
[8]   Institute of Anatomy and Cell Biology, School of Medicine, National Yang Ming University, Taipei 11221, Taiwan; incubator.lee@ym.edu.tw
[9]   Center for General Education, National Taitung University, Taitung 95092, Taiwan
[10] Department of Appiled Chemistry, Chaoyang University of Technology, Taichung 41349, Taiwan
*   Correspondence: tzengym@gmail.com (Y.-M.T.); changww@csmu.edu.tw (W.-W.C.);
     Tel.: +886-89-517-300 (Y.-M.T.); +886-4-2473-0022 (ext.12317) (W.-W.C.)

Academic Editor: Carmela Fimognari
Received: 8 March 2016; Accepted: 20 April 2016; Published: 28 April 2016

Abstract: Cancer stem/progenitor cells (CSCs) are a subpopulation of cancer cells involved in tumor initiation, resistance to therapy and metastasis. Targeting CSCs has been considered as the key for successful cancer therapy. Ovatodiolide (Ova) is a macrocyclic diterpenoid compound isolated from *Anisomeles indica* (L.) Kuntze with anti-cancer activity. Here we used two human breast cancer cell lines (AS-B145 and BT-474) to examine the effect of Ova on breast CSCs. We first discovered that Ova displayed an anti-proliferation activity in these two breast cancer cells. Ova also inhibited the self-renewal capability of breast CSCs (BCSCs) which was determined by mammosphere assay. Ova dose-dependently downregulated the expression of stemness genes, octamer-binding transcription factor 4 (Oct4) and Nanog, as well as heat shock protein 27 (Hsp27), but upregulated SMAD ubiquitin regulatory factor 2 (SMURF2) in mammosphere cells derived from AS-B145 or BT-474. Overexpression of Hsp27 or knockdown of SMURF2 in AS-B145 cells diminished the therapeutic effect of ovatodiolide in the suppression of mammosphere formation. In summary, our data reveal that Ova displays an anti-CSC activity through SMURF2-mediated downregulation of Hsp27. Ova could be further developed as an anti-CSC agent in the treatment of breast cancer.

Keywords: ovatodiolide; cancer stem/progenitor cells; Hsp27; SMURF2

1. Introduction

Cancer stem/progenitor cells (CSCs) have been described for decades and these particular cancer cells have been reported to be involved in tumor initiation, resistance to chemotherapy or radiotherapy, and metastasis [1,2]. Breast CSCs were first identified by Al-Hajj *et al.* with the marker of CD24-CD44+ [3]. Ginestier *et al.* later reported that breast cancer cells with high intracellular aldehyde dehydrogenase (ALDH) activity also represented the population of BCSCs [4]. In addition to cell surface markers or intracellular enzyme activity, BCSCs could be enriched with a cultivation method of the mammosphere, a clump of cancer cells with stem/progenitor cell properties [5]. The drug screening results from tumorsphere assay have been reported to be more translatable than those from the 2-dimensional adherent condition [6–9]. Targeting CSCs is considered as a key for successful treatment in cancer [2,10].

Heat shock proteins (Hsps) are a group of stress-induced proteins with a molecular chaperone function to maintain or correct the structure of intracellular proteins [11]. Several Hsps have been reported to be overexpressed in cancers, such as Hsp90 and Hsp27 [12]. Hsp27 belongs to small Hsps and its high expression in breast cancer tissues has been reported to be associated with lymph node metastasis [13]. We previously discovered that Hsp27 was upregulated in ALDH+ BCSCs [14]. Knockdown of Hsp27 in ALDH+ BCSCs resulted in the inhibition of epithelial-mesenchymal transition (EMT) and tumorigenicity [14]. We also demonstrated that the phosphorylation of Hsp27 was involved in the epidermal growth factor (EGF)-induced vasculogenic mimicry activity of BCSCs [15]. Agents that display the activity in Hsp27 inhibition are potentially being developed as anti-breast cancer drugs.

Ovatodiolide (Ova) is a macrocyclic diterpenoid compound extracted from *Anisomeles indica* (L.) Kuntze [16] with activities of anti-inflammation [17], anti–*Helicobacter pylori* [18], dermatological whitening [19], and anti-neoplasm [20–23]. Here we report that Ova displays an anti-CSC activity in breast cancer. Ova dose-dependently suppressed the self-renewal property of BCSCs and inhibited the expression of stemness genes, such as octamer-binding transcription factor 4 (Oct4) and Nanog. We further demonstrated that the anti-BCSC activity of Ova was mediated by the downregulation of Hsp27 through the induction of SMAD-specific E3 ubiquitin protein ligase 2 (SMURF2).

2. Results

2.1. Ovatodiolide Inhibited Self-Renewal Capability of BCSCs

We first determined the effect of Ova in cell proliferation of breast cancer cells. With the WST-1 assay, Ova displayed an anti-proliferation effect on AS-B145 and BT-474 human breast cancer cells and the IC_{50} value was 6.55 ± 0.78 µM (Figure 1A) and 4.80 ± 1.06 µM (Figure 1B) for AS-B145 and BT-474, respectively. Mammosphere cultivation is a method to enrich and to analyze the self-renewal capability of BCSCs [8]. We next applied the mammosphere assay to evaluate the anti-self-renewal activity of Ova. AS-B145 or BT-474 cells were cultivated into primary mammospheres in the presence of Ova at the concentration of 1 or 4 µM, which was below the IC_{50} value in the proliferation inhibition effect, and the self-renewal capability of primary spheres was determined by the formation of secondary mammospheres without Ova treatment. As shown in Figure 2, Ova dose-dependently inhibited the formation of the secondary mammosphere of AS-B145 (Figure 2A) and BT-474 (Figure 2B). The CD24-CD44+ BCSCs were also analyzed in AS-B145 or BT-474 sphere cells. After treatment of Ova at a concentration of 4 µM, the population of CD24-CD44+ cells in mammospheres of AS-B145 (Figure 2C) or BT-474 (Figure 2D) was decreased (from 99.8% to 48.5% for AS-B145 and from 87.1% to 29.9% for BT-474). From these results, Ova displayed an anti-self-renewal activity in BCSCs.

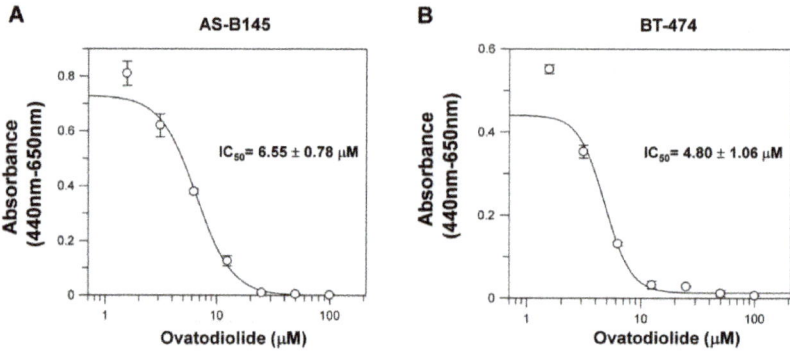

Figure 1. The cytotoxic effect of ovatodiolide in human breast cancer cells. AS-B145 (**A**) or BT-474 (**B**) cells were seeded in a 96-well plate and treated with a different concentration of ovatodiolide (0, 1.625, 3.125, 6.25, 12.5, 25, 50, 100 μM) for 72 h (*n* = 4 for each concentration). Cell proliferation was determined by WST-1 reagent and the IC50 value was calculated by GraFit software. The experiments were repeated two times and results from a representative experiment were presented.

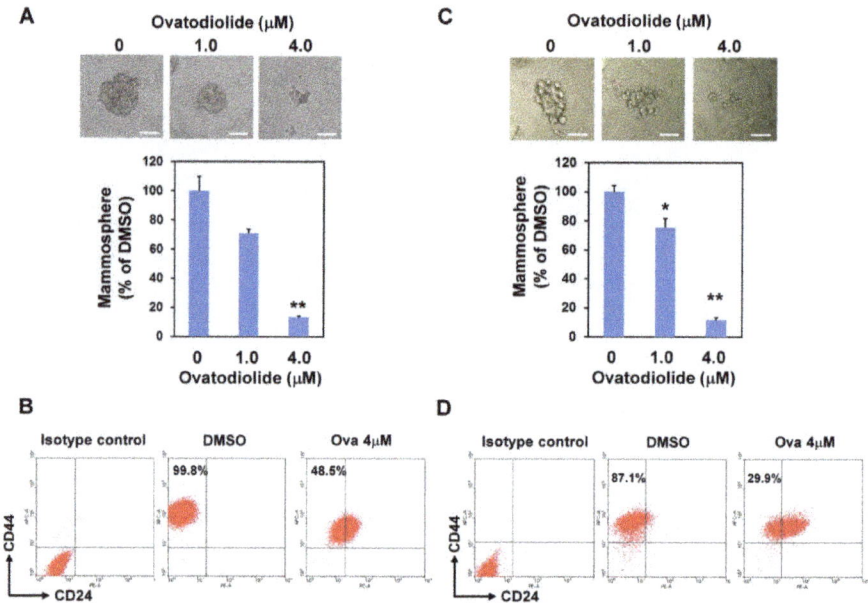

Figure 2. Ovatodiolide suppresses the self-renewal property of BCSCs. AS-B145 (**A**) or BT-474 (**B**) cells were seeded into ultralow attachment in a six-well plate under 0.1% DMSO or different concentrations of ovatodiolide (1 or 4 μM) for seven days and the formed primary mammospheres were collected and dissociated into a single cell suspension. The same number of dissociated primary sphere cells was used to evaluate the effect of ovatodiolide on the self-renewal property of BCSCs by secondary mammosphere formation without treatment of ovatodiolide (*n* = 3 for each treatment). The experiments were repeated two times and results from a representative experiment were presented. Data were presented as relative percentage of DMSO control. Scale bar = 50 μm. *, $p < 0.05$; **, $p < 0.01$. The mammosphere cells of AS-B145 (**C**) or BT-474 (**D**) were harvested and dissociated into a single-cell suspension. CD24-CD44+ cells were analyzed by flow cytometry.

2.2. Ovatodiolide Downregulated the Expression of Stemness Genes and Hsp27 but Upregulated SMURF2 Expression

We next examined the effect of Ova on the expression of stemness genes. With Western blot analysis, Ova dose-dependently inhibited the expression of Oct4 (Figure 3C) and Nanog (Figure 3D) in mammosphere cells derived from AS-B145 (Figure 3A) or BT-474 (Figure 3B) and the inhibitory effect was significantly observed at a concentration of 4 μM. We previously demonstrated that Hsp27 regulated the self-renewal and tumorigenicity of BCSCs through modulating EMT and NF-κB activity [14]. The effect of Ova in Hsp27 expression in mammosphere cells was examined. As shown in Figure 3, Ova dose-dependently downregulated Hsp27 expression (Figure 3E) in mammosphere cells derived from AS-B145 (Figure 3A) or BT-474 (Figure 3B). However, the mRNA expression of Hsp27 in mammosphere cells of AS-B145 or BT-474 was not inhibited by Ova treatment (Figure S1). A previous report indicated that SMURF2 mediated the ubiquitin-dependent degradation of Hsp27 in A549 lung cancer cells [24]. We further examined the expression of SMURF2 in mammosphere cells after Ova treatment and results revealed that Ova dose-dependently upregulated SMURF2 expression (Figure 3F) in AS-B145 (Figure 3A) or BT-474 (Figure 3B).

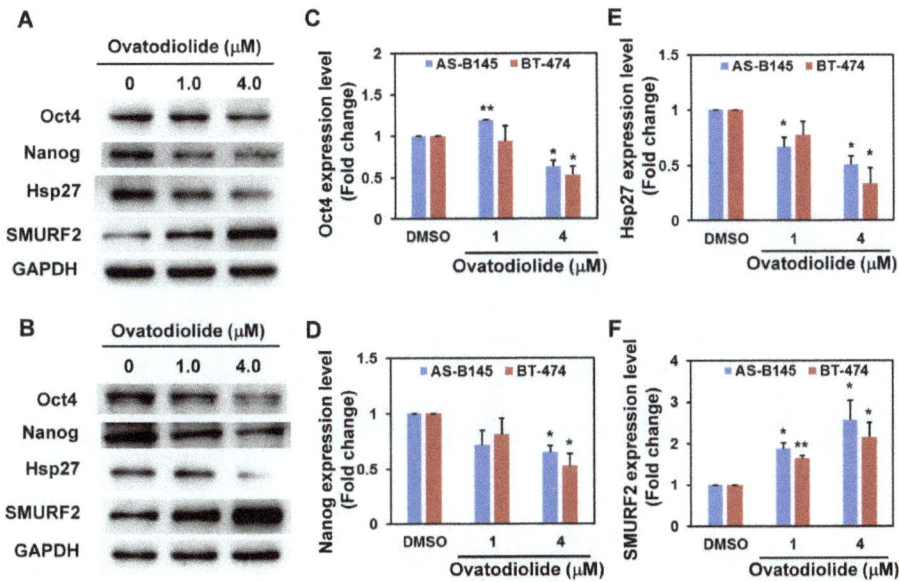

Figure 3. The change of protein expression in ovatodiolide-treated BCSCs. BCSCs were first enriched by primary mammosphere cultivation from AS-B145 (**A**) or BT-474 (**B**), dissociated into a single-cell suspension, and treated with different concentrations of ovatodiolide (0, 1, 4 μM) for 72 h (n = 2 for each treatment). The expressions of Oct4, Nanog, Hsp27 and SMURF2 were determined by Western blot. The quantification results of Oct4 (**C**), Nanog (**D**), Hsp27 (**E**) and SMURF2 (**F**) were determined by Image J software. The experiments were repeated three times and results from two representative experiments were used for quantifications. * $p < 0.05$; ** $p < 0.01$.

2.3. Overexpression of Hsp27 or Knockdown of SMURF2 Alleviated the Inhibitory Effect of Ovatodiolide

We next examined if overexpression of Hsp27 could alleviate the inhibitory effect of Ova on the mammosphere formation capability of AS-B145 cells. Overexpression of Hsp27 in AS-B145 or BT-474 cells was performed by lentivirus-mediated gene delivery and confirmed by Western blot (Figure 4A). Exogenous Hsp27 expression in AS-B145 (Figure 4B,C) or BT-474 (Figure 4D,E) significantly diminished the suppressive activity of Ova on mammosphere formation at a concentration of 4 μM ($p = 2.9 \times 10^{-6}$ for AS-B145 and $p = 1.6 \times 10^{-4}$ for BT-474). We further performed the knockdown of SMURF2 in AS-B145 or BT-474 cells by lentivirus delivery of two independent shRNA clones and both clones efficiently knocked down the mRNA expression of SMURF2 (Figure 5A). The inhibitory effect of Ova in mammosphere formation was alleviated in both sh-SMURF2 clone-transduced AS-B145 or BT-474 cells (Figure 5B,C). We also analyzed the protein expression of Hsp27 and SMURF2 in shRNA-carrying lentviruse-transduced AS-B145 or BT-474 mammosphere cells after Ova treatment. As shown in Figure 5D, these two sh-SMURF2 lentiviruses efficiently knocked down SMURF2 protein expression in AS-B145 and BT-474 mammosphere cells and alleviated the inhibitory effect of Ova in Hsp27 protein expression (Figure 5D). These results suggest the suppressive effect of Ova in the self-renewal of BCSCs through the SMURF2-mediated downregulation of Hsp27 expression.

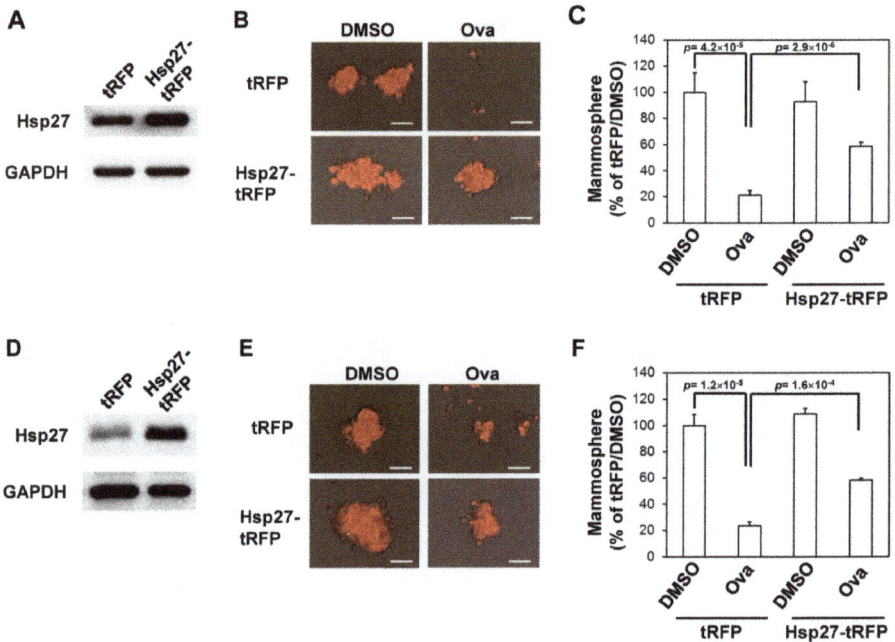

Figure 4. Overexpression of Hsp27 diminishes the inhibitory effect of ovatodiolide on the self-renewal capability of BCSCs. AS-B145 or BT-474 cells were transduced with tRFP or Hsp27-tRFP lentivirus and selected by 20 μg/mL blasticidin S for one week. The overexpression of Hsp27 was confirmed by Western blot ((**A**) for AS-B145 and (**D**) for BT-474). BCSCs were first enriched by primary mammosphere cultivation from tRFP- or Hsp27-overexpressed cells, dissociated into a single-cell suspension, and underwent secondary mammosphere formation under treatment with 4 μM ovatodiolide (Ova) or 0.1% DMSO (*n* = 3 for each treatment). Formed mammospheres were pictured ((**B**) for AS-B145 and (**E**) for BT-474) and were counted at Day 7 and displayed as the relative percentage of the DMSO group ((**C**) for AS-B145 and (**F**) for BT-474). The experiments were repeated two times and results from a representative experiment were presented. Scale bar = 50 μm.

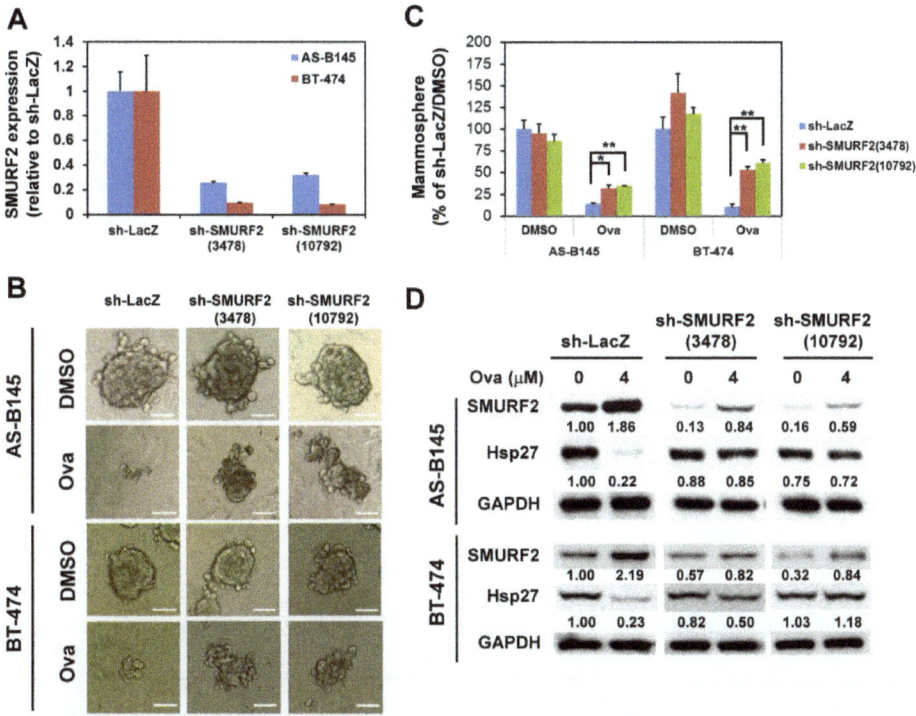

Figure 5. Knockdown of SMURF2 alleviates the suppressive effect of ovatodiolide on the self-renewal capability of BCSCs. AS-B145 or BT-474 cells were transduced with shellacs, sh-SMURF2(3478), or sh-SMURF2(10792) carrying lentivirus and selected by 2 μg/mL puromycin for three days. The knockdown efficiency was determined by qRT-PCR (**A**). After puromycin selection, the surviving cells were first cultured for primary mammosphere formation. The self-renewal capability of primary mammospheres under 4 μM ovatodiolide (Ova) or 0.1% DMSO was determined by the formation of secondary mammospheres (*n* = 3 for each treatment). Formed secondary mammospheres were pictured (**B**) and were counted at Day 7 and displayed as the relative percentage of the DMSO-treated sh-LacZ group (**C**). Scale bar = 50 μm. *, *p* < 0.05; **, *p* < 0.01. The expression of Hsp27 or SMURF2 was further determined by Western blot (**D**). The experiments were repeated two times and results from a representative experiment were presented.

3. Discussion

Recently, Bamodu *et al.* reported that Ova sensitized breast cancer cells to doxorubicin and inhibited their CSC activity [22]. Here we also demonstrated the anti-self-renewal activity of Ova in BCSCs (Figure 2). Hsp27 has been reported to be involved in the chemoresistant property of CSCs [25,26]. Knockdown of Hsp27 increased the susceptibility of Herceptin-resistant SKBR3 cells to Herceptin [27]. The survival role of Hsp27 has been reported to be mediated through the activation of Akt [28] and inhibition of Bax activation [29]. Together with our findings, the sensitization activity of Ova to chemotherapy drugs in breast cancer cells may also be mediated through the downregulation of Hsp27.

Lin *et al.* previously demonstrated that Ova inhibited the metastatic ability of MDA-MB-231 breast cancer cells through suppression of p38 mitogen-activated protein kinase (MAPK) and phosphatidylinositol 3-kinase/Akt activation [21]. p38-MAPK has been known as one of the upstream kinases in Hsp27 phosphorylation [30]. We previously reported that the activation of p38 MAPK and

Hsp27 phosphorylation was upregulated in BCSCs [14]. The knockdown of Hsp27 in BCSCs resulted in the suppression of the EMT features [14]. Hsp27 has been demonstrated to be involved in transforming growth factor β–induced EMT through the maintenance of Snail expression [31]. The emergence of CSCs could be a result of EMT [32]. Here we reported the inhibitory activity of Ova on the Hsp27 expression of BCSCs (Figure 3). The anti-CSC activity of Ova may be mediated through its inhibitory activity in EMT but remains to be further investigated. Ho *et al.* reported that Ova inhibited β-catenin signaling in renal cell carcinoma [23]. Hsp27 has been reported to interact with cytoplasmic β-catenin in breast cancer cells and clinical specimens [33]. It is worth investigating the effect of Ova on the Hsp27–β-catenin interaction in BCSCs.

4. Conclusions

In conclusion, we observed that Ova displayed anti-self-renewal activity in the mammopshere formation of human breast cancer cells as well as lead to the downregulation of stemness genes such as Oct4 and Nanog. The anti-BCSC activity of Ova was mediated through SMURF2-mediated downengulation of Hsp27. These findings reveal the potential of Ova in breast cancer treatment.

5. Materials and Methods

5.1. Reagents and Antibodies

Ovatodiolide was prepared as the previous report [17]. WST-1 cell proliferation reagent was purchased from Roche Life Science (Indianapolis, IN, USA). Polyclonal rabbit anti-Oct4 (CST-2750) or anti-Nanog (CST-3580) antibodies were purchased from Cell Signaling Technology, Inc. (Danvers, MA, USA). Polyclonal rabbit anti-Hsp27 antibody (ADI-SPA-803) was purchased from Enzo Life Sciences, Inc. (Farmingdale, NY, USA). Polyclonal rabbit anti-SMURF2 antibody (GTX110487) was purchased from GeneTex International Corporation (Hsinchu City, Taiwan). Peroxidase conjugated plo, yclonal goat anti-rabbit IgG secondary antibody (AP132P) was purchased from Merck Millipore (Temecula, CA, USA).

5.2. Cell Culture and Cytotoxicity Assay

AS-B145 human breast cancer cells were established from a female breast cancer specimen as our previous report [14,34] and maintained in MEMα (Gibco, Invitrogen Corporation, Carlsbad, CA, USA) supplemented with 10% fetal bovine serum (FBS, Gibco) and 5 µg/mL insulin (Sigma-Aldrich, St. Louis, MO, USA) at 37 °C, 5% CO_2 incubator. BT-474 human breast cancer cells were obtained from The Bioresource Collection and Research Center in Taiwan (Hsinchu City, Taiwan) and cultured in DMEM/F12 (Gibco) supplemented with 10% FBS. For determination of cytotoxic effect of Ova, AS-B145 or BT-474 cells were seeded as 2×10^4 cells/well in 96-well plates in presence of different concentration of Ova and cultured at 37 °C for 72 h. The proliferation/survival of cells was determined by adding WST-1 reagent and measured the absorbance of 440 and 650 nm wavelength. IC_{50} values were calculated by GraFit software (version 7, Erithacus Software Ltd., West Sussex, UK, 2012).

5.3. Mammosphere Cultivation

For primary mammosphere formation, AS-B145 or BT-474 cells were suspended as 1×10^4/well/2 mL and cultured in ultralow attachment 6-well-plates (Corning, Lowell, MA, USA) in DMEM/F12 medium supplemented with 0.4% bovine serum albumin (Sigma), 20 ng/mL EGF (PeproTech, Rocky Hill, NJ, USA) , 20 ng/mL basic fibroblast growth factor (Sino Biological Inc., Beijing, China), 4 µg/mL heparin (Sigma-Aldrich), 5 µg/mL insulin, 1 µM hydrocortisone (Sigma-Aldrich), and 1X B27 supplement (Gibco). After 7 days, primary mammospheres with diameter larger than 50 µm were collected with 70 µm cell strainer (BD Biosciences, San Jose, CA, USA) and dissociated into single cell suspension with HyQTase (GE Healthcare Life Sciences HyClone Laboratories, Logan, UT, USA) for secondary mammosphere formation at a cell density as 2500 cells/well/2mL for further seven days. The form secondary mammospheres were pictured and counted with an inverted microscopy (AE30, Motic Electric Group Co., Ltd., Xiamen, China).

5.4. Analysis of CD24-CD44+ Cells

Mammosphere cells were dissociated into single cell suspension with HyQTase solution. 1×10^5 dissociated cells were stained by anti-human CD24-PE (Cat. No. 555428, BD Biosciences) and anti-human CD44-APC (Cat. No. 559942, BD Biosciences) conjugated antibodies in staining buffer (PBS containing 1% FBS and 0.05% NaN$_3$) on ice for 30 min. PE conjugated mouse IgG2a κ (Cat. No. 555574, BD Biosciences) and APC conjugated mouse IgG2b κ (Cat. No. 555745, BD Biosciences) antibodies were used as isotype controls. After washing with 2 ml PBS/0.01% NaN$_3$, the fluorescence signals of stained cells were acquired with FACSAria^{TM} flow cytometer (BD Biosciences) and the data were further analyzed by WinMDI software (version 2.9, The Scripps Research Institute, La Jolla, CA, USA, 2000).

5.5. Western Blot

Mammospheres were collected by centrifugation and lysed in M-PER Mammalian Protein Extraction Reagent (Pirece Thermo Fisher Scientific Inc., Waltham, MA, USA). Then 20 μg of total protein were separated by SDS-PAGE and transferred onto PVDF membrane (Immobilon-P, Merck Millipore). The membrane was then blocked with 5% skimmed milk (Sigma-Aldrich, St. Louis, MO, USA) dissolved in Tris buffered slaine (Sigma-Aldrich) containing 0.05% Tween-20 (Sigma-Aldrich) (TBS-T) at room temperature for 1 h followed by incubation with primary antibodies at 4 °C overnight. After washing with TBS-T, the membrane was then incubated with peroxidase conjugated secondary antibodies (PerkinElmer, Waltham, MA, USA) at room temperature for one hour. The signals were developed by ECL-plus chemiluminescence substrate (PerkinElmer) and captured using a Luminescent Image Analyzer (Fusion SOLO, Vilber Lourmat, Marne-la-Vallée, France). The band intensity was quantified using ImageJ software (version 1.48a, NIH, Bethesda. MA, USA, 2013).

5.6. Lentivirus Production and Transduction

Hsp27 cDNA was amplified from pDsRed-Hsp27 [14] and cloned into pLAS5w.Pbsd-L-tRFP lentiviral plasmid (obtained from the National RNAi Core Facility at the Institute of Molecular Biology, Academia Sinica, Taipei, Taiwan) with following primers:

AfeI-Hsp27-F: 5'-AATAGCGCTATGACCGAGCGCCGCGTCCCC-3'
Hsp27-EcoRI-R: 5'-CGCGAATTCTTACTTGGCGGCAGTCTCATC-3'

The amplified DNA fragments and pLAS5w.Pbsd-L-tRFP plasmid were digested with AfeI and EcoRI restriction enzymes followed by ligation and transformation into Stbl3 competent cells (Life Technologies, Carlsbad, CA, USA). Lentiviral shRNA plasmids (TRCN00023122 for sh-LacZ, TRCN0000003478 for sh-SMURF2(3478), or TRCN0000010792 for sh-SMURF2(10792)) were also obtained from the National RNAi Core Facility. For lentivirus production, individual lentiviral plasmid was mixed with pCMV-ΔR8.91 and pMD.G plasmids as a ratio of 1:0.9:0.1 and transfected into 293T cells by GenJet^{TM} Transfection Reagent (SignaGen Laboratories, Ijamsville, MD, USA). Cells were transduced with lentivirus as MOI = 1 in presence of 8 μg/mL polybrene (Sigma-Aldrich) for 24 h and selected by blasticidin S (for Hsp27) or puromycin (for sh-LacZ or sh-SMURF2 clones) (TOKU-E Company, Bellingham, WA, USA).

5.7. Quantitative Reverse Transcription Polymerase Chain Reaction (qRT-PCR)

Total RNA was extracted by Quick-RNA^{TM} MiniPrep Kit (Zymo Research Corporation, Irvine, CA, USA). Then 1 μg of extracted RNA was used for cDNA synthesis with random hexamers provided in RevertAid First Strand cDNA Synthesis Kit (Thermo Fisher Scientific Inc., Waltham,

MA, USA) followed by SYBR Green-based qPCR reaction (SYBR® FAST qPCR Kit, Kapa Biosystems, Inc., Wilmington, MA, USA). The cycling conditions were as follows: 50 °C for 2 min, 95 °C for 10 min, followed by 40 cycles of 95 °C for 10 sec and 60 °C for 1 min. The end-point used in the real-time quantification was calculated by the StepOne software (v2.2.2, Applied Biosystmes, Carlsbad, CA, USA, 2011), and the threshold cycle number (Ct value) for each analyzed sample was calculated. The primer sets used in this study were listed as followed:

SMURF2

Forward: 5'-TAGCCCTGGCAGACCTCTTA-3'

Reverse: 5'- AATACACCTGGCCTTGTTGC-3'

MRPL19 (internal control)

Forward: 5'- GGGATTTGCATTCAGAGATCAG-3'

Reverse: 5'- GGAAGGGCATCTCGTAAG-3'

qPCR data were analyzed as previous described [35].

5.8. Statistical Analysis

Quantitative data were presented as the mean ± SD. The comparisons between two groups were analyzed with Student's *t*-test. The comparisons among multiple groups (more than three) were analyzed with one-way ANOVA and performed post-hoc test with Tukey Multiple Comparison analysis. A *p*-value of less that 0.05 was considered significantly different.

Supplementary Materials: The following are available online at www.mdpi.com/2072-6651/8/5/127/s1, Figure S1: Ovatodiolide did not inhibit Hsp27 mRNA expression in BCSCs. BCSCs were enriched by primary mammosphere cultivation from AS-B145 or BT-474 breast cancer cells and dissociated into single cell suspension followed by treated with ovatodilide (1 or 4 mM) or 0.1% DMSO for 48 h. Total RNA were than extracted and the Hsp27 mRNA expression was determined by SYBR Green-based qRT-PCR method.

Acknowledgments: This work is supported by the Ministry of Science and Technology of Taiwan (grant No. MOST 103-2113-M-324-001-MY2 (Y.-M.T.) and MOST 103-2314-B-040-015-MY3 (W.-W.C.)) and the inter-institutional research grant between Changhua Christian Hospital and Chung Medical University (grant No. CSMU-CCH-104-03 (W.-W.C. and K.-T.L.)).

Author Contributions: J.-J.Y., W.-W.C. and Y.-M.Y. conceived and designed the experiments; W.-Y.C. and J.C.-C. performed the experiments; K.-T.L., W.-Y.C. and W.-W.C. analyzed the data; K.-T.L., B.-Y.W. and Y.-M.T. contributed reagents/materials; H.-T.L., W.-W.C. and Y.-M.T. wrote the paper.

Conflicts of Interest: The authors declare no conflict of interest.

Abbreviations

The following abbreviations are used in this manuscript:

CSCs	Cancer stem/progenitor cells
Hsp27	heat shock protein 27
Ova	Ovatiodiolide
Oct4	POU class 5 homeobox 1
Nanog	Nanog homeobox
SMURF2	SMAD ubiquitin regulatory factor 2

References

1. Ajani, J.A.; Song, S.; Hochster, H.S.; Steinberg, I.B. Cancer stem cells: The promise and the potential. *Semin. Oncol.* **2015**, *42* (Suppl. 1), S3–S17. [CrossRef] [PubMed]
2. Maccalli, C.; de Maria, R. Cancer stem cells: Perspectives for therapeutic targeting. *Cancer Immunol. Immunother. CII* **2015**, *64*, 91–97. [CrossRef] [PubMed]
3. Al-Hajj, M.; Wicha, M.S.; Benito-Hernandez, A.; Morrison, S.J.; Clarke, M.F. Prospective identification of tumorigenic breast cancer cells. *Proc. Natl. Acad. Sci. USA* **2003**, *100*, 3983–3988. [CrossRef] [PubMed]

4. Ginestier, C.; Hur, M.H.; Charafe-Jauffret, E.; Monville, F.; Dutcher, J.; Brown, M.; Jacquemier, J.; Viens, P.; Kleer, C.G.; Liu, S.; *et al.* Aldh1 is a marker of normal and malignant human mammary stem cells and a predictor of poor clinical outcome. *Cell Stem Cell* **2007**, *1*, 555–567. [CrossRef] [PubMed]

5. Ponti, D.; Costa, A.; Zaffaroni, N.; Pratesi, G.; Petrangolini, G.; Coradini, D.; Pilotti, S.; Pierotti, M.A.; Daidone, M.G. Isolation and *in vitro* propagation of tumorigenic breast cancer cells with stem/progenitor cell properties. *Cancer Res.* **2005**, *65*, 5506–5511. [CrossRef]

6. Hongisto, V.; Jernstrom, S.; Fey, V.; Mpindi, J.P.; Kleivi Sahlberg, K.; Kallioniemi, O.; Perala, M. High-throughput 3d screening reveals differences in drug sensitivities between culture models of JIMT1 breast cancer cells. *PLoS ONE* **2013**, *8*, e77232. [CrossRef]

7. Kim, S.; Alexander, C.M. Tumorsphere assay provides more accurate prediction of *in vivo* responses to chemotherapeutics. *Biotechnol. Lett.* **2014**, *36*, 481–488. [CrossRef] [PubMed]

8. Lee, C.H.; Yu, C.C.; Wang, B.Y.; Chang, W.W. Tumorsphere as an effective *in vitro* platform for screening anti-cancer stem cell drugs. *Oncotarget* **2016**, *7*, 1215–1226.

9. Morrison, B.J.; Hastie, M.L.; Grewal, Y.S.; Bruce, Z.C.; Schmidt, C.; Reynolds, B.A.; Gorman, J.J.; Lopez, J.A. Proteomic comparison of mcf-7 tumoursphere and monolayer cultures. *PLoS ONE* **2012**, *7*, e52692. [CrossRef]

10. Bouvard, C.; Barefield, C.; Zhu, S. Cancer stem cells as a target population for drug discovery. *Future Med. Chem.* **2014**, *6*, 1567–1585. [CrossRef]

11. Powers, M.V.; Workman, P. Inhibitors of the heat shock response: Biology and pharmacology. *FEBS Lett.* **2007**, *581*, 3758–3769. [CrossRef]

12. Soo, E.T.; Yip, G.W.; Lwin, Z.M.; Kumar, S.D.; Bay, B.H. Heat shock proteins as novel therapeutic targets in cancer. *In Vivo* **2008**, *22*, 311–315. [PubMed]

13. Kaigorodova, E.V.; Zavyalova, M.V.; Bogatyuk, M.V.; Tarabanovskaya, N.A.; Slonimskaya, E.M.; Perelmuter, V.M. Relationship between the expression of phosphorylated heat shock protein beta-1 with lymph node metastases of breast cancer. *Cancer Biomark* **2015**, *15*, 143–150. [PubMed]

14. Wei, L.; Liu, T.T.; Wang, H.H.; Hong, H.M.; Yu, A.L.; Feng, H.P.; Chang, W.W. Hsp27 participates in the maintenance of breast cancer stem cells through regulation of epithelial-mesenchymal transition and nuclear factor-kappab. *Breast Cancer Res. BCR* **2011**, *13*, R101. [CrossRef]

15. Lee, C.H.; Wu, Y.T.; Hsieh, H.C.; Yu, Y.; Yu, A.L.; Chang, W.W. Epidermal growth factor/heat shock protein 27 pathway regulates vasculogenic mimicry activity of breast cancer stem/progenitor cells. *Biochimie* **2014**, *104*, 117–126. [CrossRef]

16. Arisawa, M.; Nimura, M.; Fujita, A.; Hayashi, T.; Morita, N.; Koshimura, S. Biological active macrocyclic diterpenoids from chinese drug "fang feng cao"; II. Derivatives of ovatodiolids and their cytotoxity. *Planta Med.* **1986**, *4*, 297–299. [CrossRef]

17. Rao, Y.K.; Chen, Y.C.; Fang, S.H.; Lai, C.H.; Geethangili, M.; Lee, C.C.; Tzeng, Y.M. Ovatodiolide inhibits the maturation of allergen-induced bone marrow-derived dendritic cells and induction of TH2 cell differentiation. *Int. Immunopharmacol.* **2013**, *17*, 617–624. [CrossRef]

18. Lien, H.M.; Wang, C.Y.; Chang, H.Y.; Huang, C.L.; Peng, M.T.; Sing, Y.T.; Chen, C.C.; Lai, C.H. Bioevaluation of anisomeles indica extracts and their inhibitory effects on helicobacter pylori-mediated inflammation. *J. Ethnopharmacol.* **2013**, *145*, 397–401. [CrossRef]

19. Huang, H.C.; Lien, H.M.; Ke, H.J.; Chang, L.L.; Chen, C.C.; Chang, T.M. Antioxidative characteristics of anisomeles indica extract and inhibitory effect of ovatodiolide on melanogenesis. *Int. J. Mol. Sci.* **2012**, *13*, 6220–6235. [CrossRef] [PubMed]

20. Liao, Y.F.; Rao, Y.K.; Tzeng, Y.M. Aqueous extract of anisomeles indica and its purified compound exerts anti-metastatic activity through inhibition of nf-kappab/ap-1-dependent MMP-9 activation in human breast cancer MCF-7 cells. *Food Chem. Toxicol.* **2012**, *50*, 2930–2936. [CrossRef] [PubMed]

21. Lin, K.L.; Tsai, P.C.; Hsieh, C.Y.; Chang, L.S.; Lin, S.R. Antimetastatic effect and mechanism of ovatodiolide in MDA-MB-231 human breast cancer cells. *Chemico-Biol. Interact.* **2011**, *194*, 148–158. [CrossRef] [PubMed]

22. Bamodu, O.A.; Huang, W.C.; Tzeng, D.T.; Wu, A.; Wang, L.S.; Yeh, C.T.; Chao, T.Y. Ovatodiolide sensitizes aggressive breast cancer cells to doxorubicin, eliminates their cancer stem cell-like phenotype, and reduces doxorubicin-associated toxicity. *Cancer Lett.* **2015**, *364*, 125–134. [CrossRef] [PubMed]

23. Ho, J.Y.; Hsu, R.J.; Wu, C.L.; Chang, W.L.; Cha, T.L.; Yu, D.S.; Yu, C.P. Ovatodiolide targets β-catenin signaling in suppressing tumorigenesis and overcoming drug resistance in renal cell carcinoma. *Evid. Based Complement. Altern. Med.* **2013**, *2013*, 161628. [CrossRef] [PubMed]

24. Sun, Y.; Zhou, M.; Fu, D.; Xu, B.; Fang, T.; Ma, Y.; Chen, J.; Zhang, J. Ubiquitination of heat shock protein 27 is mediated by its interaction with smad ubiquitination regulatory factor 2 in a549 cells. *Exp. Lung Res.* **2011**, *37*, 568–573. [CrossRef]

25. Lee, C.H.; Hong, H.M.; Chang, Y.Y.; Chang, W.W. Inhibition of heat shock protein (hsp) 27 potentiates the suppressive effect of hsp90 inhibitors in targeting breast cancer stem-like cells. *Biochimie* **2012**, *94*, 1382–1389. [CrossRef]

26. Hsu, H.S.; Lin, J.H.; Huang, W.C.; Hsu, T.W.; Su, K.; Chiou, S.H.; Tsai, Y.T.; Hung, S.C. Chemoresistance of lung cancer stemlike cells depends on activation of hsp27. *Cancer* **2011**, *117*, 1516–1528. [CrossRef]

27. Kang, S.H.; Kang, K.W.; Kim, K.H.; Kwon, B.; Kim, S.K.; Lee, H.Y.; Kong, S.Y.; Lee, E.S.; Jang, S.G.; Yoo, B.C. Upregulated hsp27 in human breast cancer cells reduces herceptin susceptibility by increasing her2 protein stability. *BMC Cancer* **2008**, *8*, 286. [CrossRef]

28. Wu, R.; Kausar, H.; Johnson, P.; Montoya-Durango, D.E.; Merchant, M.; Rane, M.J. Hsp27 regulates akt activation and polymorphonuclear leukocyte apoptosis by scaffolding MK2 to akt signal complex. *J. Biol. Chem.* **2007**, *282*, 21598–21608. [CrossRef]

29. Havasi, A.; Li, Z.; Wang, Z.; Martin, J.L.; Botla, V.; Ruchalski, K.; Schwartz, J.H.; Borkan, S.C. Hsp27 inhibits bax activation and apoptosis via a phosphatidylinositol 3-kinase-dependent mechanism. *J. Biol. Chem.* **2008**, *283*, 12305–12313. [CrossRef]

30. Kostenko, S.; Moens, U. Heat shock protein 27 phosphorylation: Kinases, phosphatases, functions and pathology. *Cell. Mol. Life Sci. CMLS* **2009**, *66*, 3289–3307. [CrossRef]

31. Wettstein, G.; Bellaye, P.S.; Kolb, M.; Hammann, A.; Crestani, B.; Soler, P.; Marchal-Somme, J.; Hazoume, A.; Gauldie, J.; Gunther, A.; et al. Inhibition of hsp27 blocks fibrosis development and emt features by promoting snail degradation. *FASEB J.* **2013**, *27*, 1549–1560. [CrossRef]

32. Singh, A.; Settleman, J. Emt, cancer stem cells and drug resistance: An emerging axis of evil in the war on cancer. *Oncogene* **2010**, *29*, 4741–4751. [CrossRef]

33. Fanelli, M.A.; Montt-Guevara, M.; Diblasi, A.M.; Gago, F.E.; Tello, O.; Cuello-Carrion, F.D.; Callegari, E.; Bausero, M.A.; Ciocca, D.R. P-cadherin and beta-catenin are useful prognostic markers in breast cancer patients; beta-catenin interacts with heat shock protein hsp27. *Cell Stress Chaperones* **2008**, *13*, 207–220. [CrossRef] [PubMed]

34. Chang, W.W.; Lee, C.H.; Lee, P.; Lin, J.; Hsu, C.W.; Hung, J.T.; Lin, J.J.; Yu, J.C.; Shao, L.E.; Yu, J.; et al. Expression of globo H and SSEA3 in breast cancer stem cells and the involvement of fucosyl transferases 1 and 2 in globo H synthesis. *Proc. Natl. Acad. Sci. USA* **2008**, *105*, 11667–11672. [CrossRef] [PubMed]

35. Chang, Y.C.; Lin, C.W.; Yu, C.C.; Wang, B.Y.; Huang, Y.H.; Hsieh, Y.C.; Kuo, Y.L.; Chang, W.W. Resveratrol suppresses myofibroblast activity of human buccal mucosal fibroblasts through the epigenetic inhibition of zeb1 expression. *Oncotarget* **2016**. [CrossRef]

toxins

MDPI

Review

Ellagitannins in Cancer Chemoprevention and Therapy

Tariq Ismail [1], Cinzia Calcabrini [2,3], Anna Rita Diaz [2], Carmela Fimognari [3], Eleonora Turrini [3], Elena Catanzaro [3], Saeed Akhtar [1] and Piero Sestili [2,*]

[1]   Institute of Food Science & Nutrition, Faculty of Agricultural Sciences and Technology,
    Bahauddin Zakariya University, Bosan Road, Multan 60800, Punjab, Pakistan;
    ammarbintariq@yahoo.com (T.I.); saeedbzu@yahoo.com (S.A.)
[2]   Department of Biomolecular Sciences, University of Urbino Carlo Bo, Via I Maggetti 26, 61029 Urbino (PU),
    Italy; anna.diaz@uniurb.it
[3]   Department for Life Quality Studies, Alma Mater Studiorum-University of Bologna, Corso d'Augusto 237,
    47921 Rimini (RN), Italy; cinzia.calcabrini@unibo.it (C.C.); carmela.fimognari@unibo.it (C.F.);
    eleonora.turrini@unibo.it (E.T.); elena.catanzaro2@unibo.it (E.C.)
*   Correspondence: piero.sestili@uniurb.it; Tel.: +39-(0)-722-303-414

Academic Editor: Jia-You Fang
Received: 31 March 2016; Accepted: 9 May 2016; Published: 13 May 2016

Abstract: It is universally accepted that diets rich in fruit and vegetables lead to reduction in the risk of common forms of cancer and are useful in cancer prevention. Indeed edible vegetables and fruits contain a wide variety of phytochemicals with proven antioxidant, anti-carcinogenic, and chemopreventive activity; moreover, some of these phytochemicals also display direct antiproliferative activity towards tumor cells, with the additional advantage of high tolerability and low toxicity. The most important dietary phytochemicals are isothiocyanates, ellagitannins (ET), polyphenols, indoles, flavonoids, retinoids, tocopherols. Among this very wide panel of compounds, ET represent an important class of phytochemicals which are being increasingly investigated for their chemopreventive and anticancer activities. This article reviews the chemistry, the dietary sources, the pharmacokinetics, the evidence on chemopreventive efficacy and the anticancer activity of ET with regard to the most sensitive tumors, as well as the mechanisms underlying their clinically-valuable properties.

Keywords: ellagitannins; phytochemicals; cancer; chemoprevention; cancer therapy; safety

1. Introduction

Despite the enormous efforts of the scientific and medical community, cancer still represents the second leading cause of death and is nearly becoming the leading one in the elderly [1]. It is estimated that, due to demographic changes alone, in the next 15 years the number of new cancer cases will increase by 70% worldwide [2].

The lack of effective diagnostic tools for early detection of several tumors, the limited treatment options for patients with advanced stages of cancer, and the onset of multiple drug resistance favor poor prognosis and high mortality rates. The significant, but still unsatisfactory, improvement of survival, the severe toxicity profile, and the high costs characterizing many current anticancer therapies clearly show that a threshold in terms of clinical benefit and patients' tolerance has been reached. Thus, the identification and development of innovative, preventive as well as therapeutic strategies to contrast cancer-associated morbidity and mortality are urgently needed.

Epidemiological, preclinical, and clinical studies have generally concluded that a diet rich in phytochemicals can reduce the risk of cancer [2,3]. Due to their pleiotropism which includes antioxidant, anti-inflammatory, and antiproliferative activities as well as modulatory effects on subcellular signaling pathways, phytochemicals from edible fruits and vegetables are recognized as an effective option to counteract cancer incidence and mortality [3–5]. Plants constitute a primary and large source of various chemical compounds including alkaloids, flavonoids, phenolics, tannins, tocopherols, triterpenes, and isothiocyanates. Ellagitannins (ET) are an important class of phytochemicals contained in a number of edible plants and fruits recommended by the traditional medicine of a variety of cultures, both in the developing and developed countries, to treat common health problems. ET biological and nutraceutical potential has received increasing attention over the last several decades. ET exert multiple and clinically-valuable activities [4], and among them the chemopreventive, anticarcinogenic, and antiproliferative activities are being receiving growing interest and attention (Figure 1).

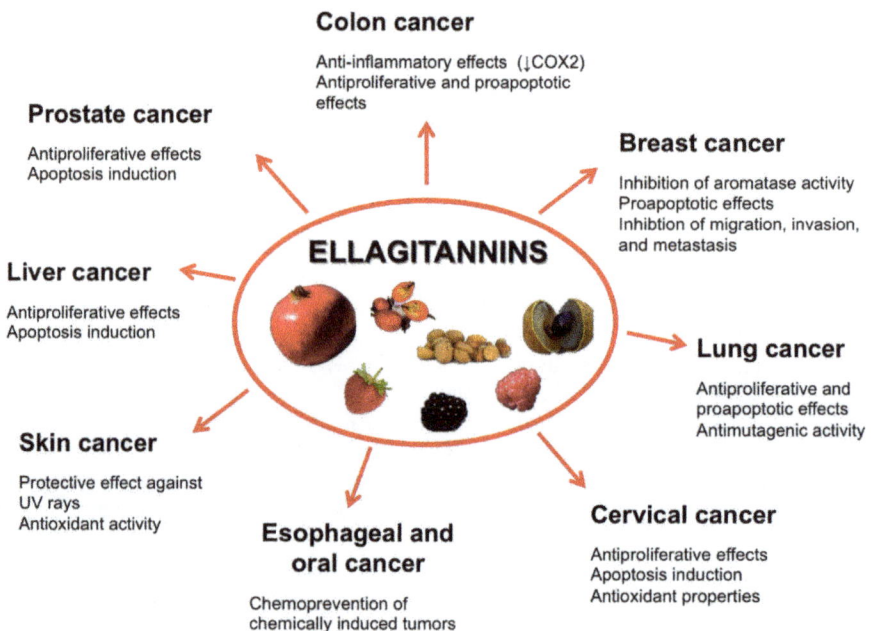

Figure 1. Fruits containing ET with chemopreventive, anticarcinogenic, and antiproliferative activities.

2. Dietary Sources, Types, and Occurrence

ET and their derivatives are noticeably contained in edible seeds, nuts, and various fruits of nutritional interests. The structures of relevant ETs and of ellagic acid are shown in Figure 2. A wide variety of fresh fruits including berries, like raspberries, black raspberries, strawberries, pomegranate, longan, and dried nuts, are renowned for their ample polyphenols concentration in the form of ET [5]. Five species of berries including raspberry, strawberry, cloudberry, rose hip, and sea buckthorn were identified by Koponen *et al.*, [6] as significant carrier of ET in a range of 1–330 mg per 100 g of fruit. Sanguiin H-6 and lambertianin C were reported from Glen Ample raspberries and Scottish-grown red raspberries, along with some trace levels of ellagic acid [7,8]. Blackberries (fruit and seeds) have been reported for a range of ET including pedunculagin, casuarictin, sanguiin H-6 (lambertianin A), and lambertianin (C and D) [9–11]. Pomegranate and various fractions of the fruit are known for their cancer chemopreventive properties owing to their unique phenolics composition in the form of ET, which include punicalagin, punicalin, granatin A, granatin B, tellimagrandin I, pedunculagin, corilagin, gallagic acid, ellagic acid, and casuarinin [12].

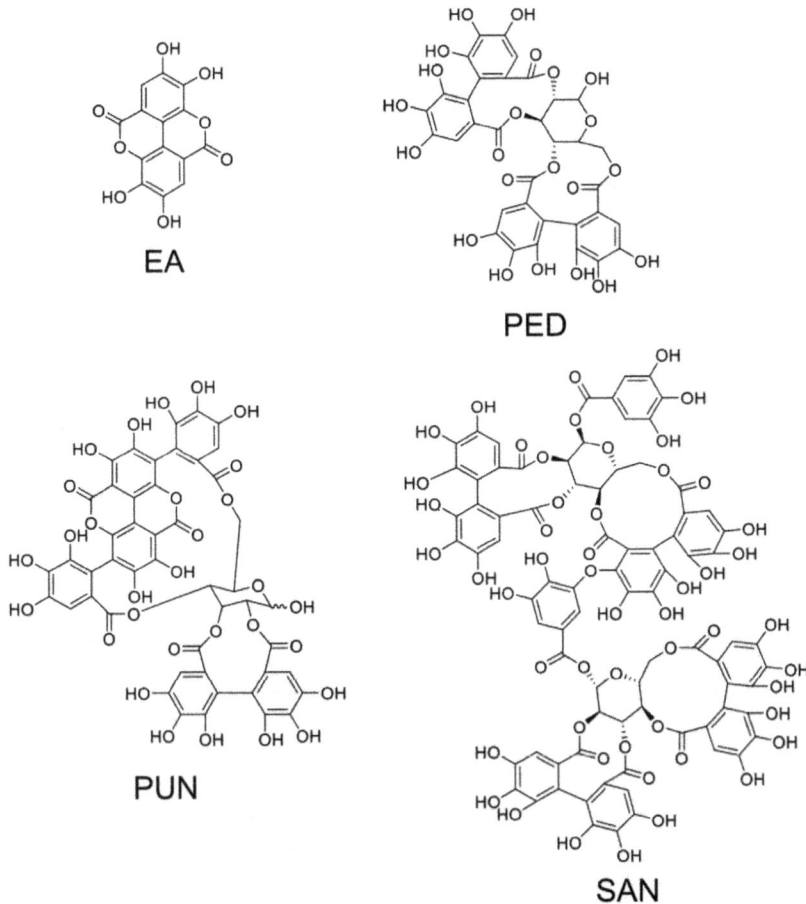

Figure 2. Structures of ellagic acid and of three major and representative ellagitannins. EA, ellagic acid; PED, pedunculagin; PUN, punicalagin; SAN, sanguiin H-6.

ET, predominately those isolated from pomegranate (e.g., punicalagin), have gained a wide popularity as preventive and therapeutic ethnopharmacological approaches for cancer treatment. However, a lot more has been added to this class of compounds from fruits other than pomegranate, including raspberries, blueberries, strawberries, muscadine grapes, and longan [7,13–18]. Major phenolic fractions recovered from longan include gallic acid, ellagic acid, and corilagin, much more concentrated in the seed segment as compared to the fruit pulp and peel [17]. Good essential fatty acid composition of nuts and fairly high concentrations of ET and their derived fractions, such as ellagic acid and its glycosidic derivatives have been associated with the potential cardioprotective properties of nuts. Ellagic acid (free and total) has been reported in a range of 0.37–823 mg per 100 g of dried nuts [19]. High concentrations of a variety of ET (ellagic acid, sanguiin H2 and 6, lambertianin C, castalagin/vescalagin, galloyl-bis-HHDP glucose, pedunculagin) can be found in blackberries (*Rubus* sp.) [20]. Shi *et al.*, [21] identified agrimoniin as the second highest phenolic compound of strawberries.

Irrespective of the edible fractions of fruiting plants, some inedible fractions like fruit peels, bark and foliage have also been reported as good source of hydrolysable tannins including bioactive ET [4,22]. Leaves extracts of *Shepherdia argentea*—a deciduous shrub commonly known as silver buffaloberry—were reported as a good reserve of gluconic acid core carrying the potential anti-HIV novel ET, such as hippophaenin A, shephagenin A and shephagenin B [23].

3. Ellagitannins—Classification and Chemistry

Tannins are unique secondary metabolites of plant phenolics with relatively higher molecular weight (300–30,000 Da) and bear the ability to generate complexes with some macromolecules, like proteins and carbohydrates [24]. Chemistry and nomenclature of the tannins is complicated by virtue of the frequent changes which parallel the advancement in this very specific field [25]. Taking into account different definitions of tannins [26,27], these compounds may be referred as either galloyl esters and their derivatives (ET, gallotannins, and complex tannins), or the oligomeric and polymeric proanthocyanidins (condensed tannins). In a broader perspective, tannins may be classified most satisfactorily and unambiguously on the basis of structural configuration and/or solubility [28]. C–C coupling of galloyl units in absence of glycosidically-linked catechin make ETs structurally different from the condensed tannins that are characterized by monomeric catechin linkages (C4–C8 or C4–C6) to generate oligomeric likewise polymeric proanthocyanidins [27]. Gallotannins and ETs constitute a major group of tannins *i.e.*, hydrolysable tannins that are well known for their properties to hydrolyze into hexahydroxydiphenol (HHDP) or gallic acid moieties. Gallotannins are the gallic acid derivatives carrying $\geqslant$ six gallyol groups and might further be characterized on account of one or more than one digalloyl group [29].

ETs (hydrolysable tannins) on their hydrolysis yield gallic acid and ellagic acid from the compounds carrying gallyol groups and HHDP groups, respectively [28]. *In vitro* digestion models declare ETs to remain stable under the normal physiological condition of the stomach [30]. However, ETs hydrolysis to free ellagic acid or their degradation may proceed in the small intestine at neutral to alkaline pH [31]. Biologically, condensed tannins and gallotannins are thought to deliver relatively higher protein precipitation properties as compare to the ETs and hence are considered potential antinutritional compounds from the class of plants polyphenolics [32]. Gallotannins and condensed tannins have also been reported as oxidatively least active tannins as compared to the ETs and on the same time gallotannins and condensed tannins have also been found to reduce pro-oxidant properties of ETs [33,34].

3.1. Simple Ellagitannins

ET (M.W. 300–20,000 Da) are non-nitrogenous compounds with at least two C–C coupled galloyl units with no glycosidically-bonded catechin unit [3,35]. ET are derivatives of 1,2,3,4,6-penta-*O*-galloyl-β-D-glucopyranose (PGG). Structurally, ET are esters of carbohydrates and or cyclitols and also include metabolic compounds derived from oxidative cleavage of either condensed or hydrolysable tannins [27,35,36]. The presence of hexahydroxydiphenol (HHDP) in a glucopyranose ring in addition to acyl units and certain HHDP metabolites such as dehydrohexahydroxydiphenol (DHHDP), valoneoyl and chebuloyl groups constitute simple ET. Tellimagrandin I and II, pedunulagin, casuarictin, and chebulagic acid originate from the specific orientation and number of acyl groups on glucose units. Variation in HHDP group originates by linking (C–C or C–O) one or more galloyl groups to HHDP unit.

Structural diversity of ET has been reported to correlate with their carrier-plants' taxonomy and evolutionary hierarchy [37]. More often, monomeric ET or oligomeric ET constitute the major tannic component of plant species. The monomeric compounds of the group include tellimagrandins I and II, pedunculagin, casuarictin, and potentillin. Type I hydrolysable tannins (*i.e.*, gallotannins) carrying HHDP in stable conformation at either the 2,3 or 4,6 position on a D-glucopyranose may be referred to as a simple ET [38–40]. Geraniin, a type III ET, is another example of monomeric simple ET carrying a DHHDP unit linked to D-gluopyranose of HHDP unit via $^1C_4$ conformation. Dimers of ET are generated by intermolecular oxidative coupling/condensation of simple ET.

3.2. Glycosidic Ellagitannins

Chemically, the C-glycosidic linkage of ET is established via intermolecular bonds between two monomeric units, one carrying anomeric carbon while the second one galloyl or HHDP group [3,35,41]. Most recently C-glycosidic ET including granadinin, vescalagin, methylvescalagin, castalagin, stachyurin, and casuarinin have been reported from the peel and seed fraction of camu-camu, a fruiting tree of Amazon rainforest [42]. Woody fractions of various fruits, particularly the nuts and berries, have also been observed to hold novel C-glycosidic ET (e.g., castalagin and vescalagin). Castacrenins D and F are two other forms of C-glycosidic ET isolated from the woody fraction of Japanese chestnut and carry gallic acid/ellagic acid moieties [43]. Treating vescalagin with *Lentinula edodes* generates quercusnins A and B that may be referred as fungal metabolites of C-glycosidic ET [44]. Castacrenins D and F isolated from chestnut wood may generate oxidative metabolites, namely castacrenins E and G, by replacing pyrogallol rings of C-glycosidic ET with cyclopentone rings [43]. Rhoipteleanins H, I, and J were reported as novel C-glycosidic ET isolated from the fruit and bark fractions of *Rhoiptelea chiliantha*. Structural configuration of rhoipteleanins H revealed the presence of cyclopentenone carboxy moieties that are generated by oxidation and rearrangement of C-glycosidic ET aromatic ring [45].

Condensate of C-glycosidic ET is another subclass of hydrolysable tannins, which includes rhoipteleanin J produced by the intermolecular condensation (C–C or C–O) of monomeric C-glycosidic ET followed by oxidation of aromatic rings of ET [45]. Wine aged in oak wood barrels is often reported to carry oak ET, particularly the condensation products of monomeric C-glycosidic ET. The studies infer C-glycosidic ET to play a significant role in modulation of organoleptic features of wine aged in oak wood barrels [46].

4. Ellagitannins Pharmacokinetics

A precise knowledge of phytochemicals' pharmacokinetics is very important to exploit their health benefits, as well as the effects of their metabolites [47]. *In vivo*, ET, instead of being absorbed directly into the blood stream, are physiologically hydrolyzed to ellagic acid, which is further metabolized to biologically-active and bioavailable derivatives, *i.e.*, urolithins, by the activity of microbiota in gastrointestinal (GI) tract [5,48]. The biological properties of ET, such as free radical scavenging, further depend on their metabolic transformation inside gut. ET recovered from pomegranate juice may be metabolically converted by gut microbiota to urolithin A, B, C, D, 8-*O*-methylurolithin A, 8,9-di-*O*-methylurolithin C, and 8,9-di-*O*-methylurolithin D, and some of these metabolites display higher antioxidant activity than the parental tannins themselves. For instance, urolithin C and D show an antioxidant capacity—as determined in a cell-based assay—which is 10- to 50-fold higher as compared to punicalagin, punicalin, ellagic acid, and gallic acid [49]. This finding suggests that intestinal transformation products of ET are likely to play a central role for the antioxidant properties at least inside the GI tract. Significant differences in urolithins' profiles in individual human subjects feed on raspberries—a renowned source of ET—have been attributed to gut microflora, whose variations on an inter-individual basis affect their capacity of hydrolyzing ET and subsequent metabolite synthesis [48,50]. The interaction of gut microbiota composition and the host endogenous excretory system is also likely to play a further role in the observed inter-individual variability [51]. ET are highly stable under the acidic environment of stomach, and retain their composition without being hydrolyzed to simpler compounds when exposed to various gastric enzymes. Consequently the complex structure of ET impedes their gastric absorption: however, the stomach might serve as the first site of absorption of free ellagic acid and pre-hydrolyzed forms of ET.

Contrary to stomach, the neutral or alkaline environments of duodenum and small intestines, characterized by pH values ranging from 7.1 to 8.4, allow ET hydrolyzation [31,41]. In humans, ET are rapidly absorbed and metabolized, as documented by [18,52]: following ingestion of pomegranate juice (at a dose containing 25 mg of ellagic acid and 318 mg of ET), ellagic acid can be found in plasma for up to 4 h while, at later times, it is no more detectable. In contrast, another study reported that no ellagic acid could be detected in plasma during the 4 h following the juice intake [53], a discrepancy which has been attributed to inter-individual variability [54]. Ellagic acid is converted by catechol-*O*-methyl transferase to dimethylellagic acid, which is then glucuronidated and excreted [52].

Finally, the microbiologically metabolized fraction of ET, *i.e.*, urolithins, is further incorporated to enterohepatic circulation system [18,53,55,56].

5. Ellagitannins for Tumor Chemoprevention and Therapy

The development of novel mechanism-based chemopreventive and antitumor approaches to fight cancer through the use of dietary substances which humans can easily accept has become an important goal. Along this line, ET have received increasing attention over the last two decades.

Similarly to other anticancer phytochemicals, ET display chemopreventive and chemotherapeutic activities [3]. The chemopreventive activity of ET and derivatives, such as ellagic acid has been primarily associated with their antioxidant capacity, that varies with the degree of hydroxylation [5,57] and depends from both a direct radical scavenging and iron chelation activity.

The well-known anti-inflammatory capacity represents another important feature of ET chemopreventive and antitumor activity [55], that being persisting inflammation involved both as a causative and a facilitating factor in carcinogenesis and cancer development [58]. For example, pomegranate ET inhibit pro-inflammatory pathways including, but not limited to, the NF-κB pathway, whose activation leads to immune reactions, inflammation, and the transcription of genes involved in cell survival, such as Bclx and inhibitors of apoptosis. Constitutive activation of NF-κB has been observed in prostate cancers, where it sustains chronic inflammation and promotes the development of high-grade prostate cancer. With respect to inflammation, it is worth noting that, similarly to many polyphenols, the antioxidant activity of ET participates to an "anti-inflammatory loop" with other mechanisms, since it lowers the levels of radicals which otherwise would act as pro-inflammatory stimuli.

The direct antiproliferative effects of ET have been attributed to multiple mechanisms (see the next subchapters) including the cell cycle arrest capacity and the properties enabling cancer cells to follow apoptosis through the mitochondrial route and self-destruction after replication [59–62]. In addition to directly targeting tumor cell survival, the cytotoxic/cytostatic activities of ET might also concur with the chemopreventive potential, since they prevent tumor cells from converting into more malignant phenotypes and from replicating.

A study on 1,3-di-*O*-galloyl-4,6-(s)-HHDP-β-D-glucopyranose (an ET from *Balanophora japonica* MAKINO) points to the complexity and multiplicity of the mechanisms contributing to the anticancer activity of ET, *i.e.*, the same complexity and multiplicity characterizing also other classes of phytochemicals. Indeed, the antiproliferative activity of 1,3-di-*O*-galloyl-4,6-(s)-HHDP-β-D-glucopyranose in human Hep-G2 liver cancer cells was also associated to an altered regulation of 25 miRNAs including the let-7 family members miR-370, miR-373, and miR-526b, identified as likely targets with roles in cell proliferation and differentiation [63]. The fact that in cell culture systems combinations of ET or of ET and other phytochemicals present in plant or fruit extracts are more cytotoxic than any single ET [64], is suggestive of the multifactorial effects, chemical synergy, and multiplicity of the mechanisms behind their antitumor activity. To this regard, the capacity of some ET to inhibit angiogenesis, a fundamental event accompanying tumor growth, both in *in vitro* and *in vivo* prostate cancer models [65], and to reduce endothelial cell growth through binding to vascular endothelial growth factor receptors [66] represents a further and significant antitumor mechanism.

In analogy to other polyphenols, ET could also be utilized to increase the sensitivity of tumor cells to standard chemotherapeutic drugs [67], with the aim of obtaining an increase of their antitumor efficacy along with a reduction of their doses and, consequently, of their severe adverse effects which often represent a limiting factor for the prosecution of the therapeutic regimens.

As a premise to the literature data discussed in the next paragraphs, it is important noting that, since ET are not absorbed systemically after oral administration as such [48], the studies where ET extracts were given to cultured cancer cells are unlikely to be predictive of the effects which could be attained after oral ingestion *in vivo*. Rather these data could be representative of intravenously administered ET, but the toxicology of this administration route is not known.

The next sections of the review will discuss more in depth the ET anticancer mechanisms and properties emerging from *in vivo* and *in vitro* studies on a panel of tumors or tumor cells which appear as potentially sensitive targets for these phytochemicals.

5.1. Prostate Cancer

Prostate cancer is the second leading cancer-associated death risk factor among U.S. males [68]. Phytochemicals originating from various food sources slow down the progression of prostate cancer, whereas a majority of other nutrients are reported to be non-effective in either preventing or curing prostate cancer [69]. Evidence-based findings support the consolidated role of fruits, vegetables, and various culinary herbs of different cultures in averting various forms of cancers, but relatively weak and inconsistent relationships have been presented so far for prostate cancer [70,71]. Somehow more promising seem to be the edible fruits containing high amounts of ET, which have been extensively tested *in vivo* for their prostate cancer inhibitory properties. As it has been shown in animal models, higher concentrations of ET are recorded in prostate and colon tissues as compared to the others [72]. Pomegranate holds one of the highest concentration of ET [55]. Antitumor activities of pomegranate fruit juice, peel extracts, and seed oil have been reported against prostate cancer cells [73]. Dose-dependent anti-proliferative and pro-apoptotic effects of pomegranate fruit extracts (10–100 μg/mL) have been documented against aggressive human prostate cancer cells (PC3) [74]: induction of pro-apoptotic mediators (Bax and Bak), downregulation of Bcl-2 and Bcl-XL, and reduced expression of cyclin-dependent kinases 2, 4, 6, and cyclins D1, D2, and E have been identified as the mechanisms responsible for these effects.

Pomegranate extract inhibited proliferation of endothelial (HUVEC) and prostate (LNCaP) cancer cells; the extract also reduced LNCaP prostate cancer xenograft size, tumor vessel density, VEGF peptide levels and HIF-α expression after four weeks of treatment in severe combined immunodeficient mice [65].

Oenothein B, a macrocyclic ET, and quercetin-3-*O*-glucuronide from *Epilobium* sp. herbs—used in traditional medicine to treat benign prostatic hyperplasia and prostatic adenoma—have been proven to strongly inhibit the proliferation of human prostate cancer cells [75]. *Hibiscus sabdariffa* leaf extracts, which contain high amounts of ellagic acid, have been reported to inhibit the growth [76] and the expressions of metastasis-related molecular proteins [75] of LNCaP cells via activation of the mitochondrial pathway and suppression of the Akt/NF-kB signaling pathway, respectively. *Terminalia chebula*—a common ayurvedic ethnic drug of the Indian subcontinent—has been recognized for its potential biological and pharmacological properties [77]. Chebulinic acid is the predominant and more characteristic ET among the various constituents of chebula fruit (*T. chebula*). Methanolic extract (70%) of *T. chebula* fruits was shown to inhibit proliferation and induce cell death in PC3 prostate cancer cells as well as in PNT1A non-tumorigenic human prostate cells in a dose-dependent manner [78]. At low concentrations, the extract promoted initiation of apoptotic cell death, while at higher doses necrosis was the predominant type of cell demise; chebulinic acid, tannic acid, and ellagic acid were the most cytotoxic phenolics, and are likely responsible for the antitumor activity of *T. chebula* fruit extracts [78].

5.2. Colon Cancer

Cancer statistics, as reported from Centers for Disease Control and Prevention, rate colon cancer as the fourth highest death factor in USA [68]. It is widely accepted that herbal sources may provide therapeutically relevant compounds for the management of colorectal cancers. In this regard, it is worth noting that World Health Organization estimates over 80% of the entire world population rely on biomolecules with broad ethnopharmacological properties as a primary health care solution [79].

The strict correlation between chronic inflammation, malignant transformation and development of colorectal cancer is widely recognized [80]. Indeed, non-steroidal anti-inflammatory drugs have proven to be effective in preventing the formation of colorectal tumors and their malignant transformation in both preclinical and clinical studies [81]. However, unwanted, sometimes severe or even fatal, side effects (ulceration, renal toxicity, gastric bleeding) represent a major limitation for the use of these synthetic drugs: in search of alternative therapeutic options, exploration and utilization of natural biomolecules as anti-inflammatory formulations are in progress [82]. Various phytochemicals modulate inflammatory cell signaling in colon cancer: among them, pomegranate ET (*i.e.*, punicalagin and ellagic acid) have been shown to suppress cyclooxygenase-2 (COX-2) protein expression in human colon cancer (HT-29) cells [83]. Exposing HT-29 cells to 50 mg/L of powdered pomegranate juice, total pomegranate ET, or punicalagin reduces the expression of COX-2 protein by 79%, 55%, and 48%, respectively, and inhibits production of pro-inflammatory prostaglandins [83]. Another study conducted by Kasimsetty *et al.*, [84] reported that pomegranate ET and their metabolites, *i.e.*, urolithins A and C, inhibit HT-29 cells proliferation via G0/G1 and G2/M arrest, followed by induction of apoptosis. Interestingly, urolithins display advantageous pharmacokinetics over other agents, in that they tend to persist in the colon through enterohepatic circulation. Scarce information is available on the mechanistic role of ET and their metabolites, mainly urolithins, in colon cancer chemoprevention. Sharma *et al.*, [85] showed that fruit ET and their metabolites inhibit canonical Wnt signaling pathway, which is involved in the development of the majority (~90%) of colon cancers. In this light, ET and their colon-derived metabolites may be most relevant in relation to cancer prevention rather than treatment. To this regard, it is important noting that the concentrations of ET and their metabolites such as ellagic acid or urolithin A resulting in a 50% inhibition of Wnt signaling in 293T human colon cancer cells are comparable with those nutritionally attainable after regular consumption of ET-rich fruits or beverages [85].

5.3. Breast Cancer

Breast cancer is the most prevalent, spontaneous hormone-associated malignancy in women and is the most common gender-related cause of death around the globe [86,87]. Estrogen is the major stimulating factor of breast cancer cells' proliferation and tumor cells' growth. Upregulation of growth hormone receptors in breast malignant cells, as compared to the normal breast tissue, points to the key role of the pituitary, as well as the growth hormones, in the development of breast cancer in humans [88].

Complementary and alternative medicines in the form of bioactive fractions and raw decoctions of herbs, edible and inedible segments of various fruits and vegetables, are under assessment for their potential in treating breast cancer [89]. Pomegranate, its juice, and other fractions of the fruit are the richest source of high-molecular-weight ET, in particular punicalagin, as compared to any other known and commonly-consumed fruit [55]. Estrogen-induced expression of peptides growth factors is the major concern in the development and growth of estrogen-responsive mammary cancer: inhibition of this circuitry is the rationale for the use of antiestrogens and aromatase inhibitors to treat these types of breast cancer [90,91]. Pomegranate ET-derived compounds have been shown to block endogenous estrogen synthesis by inhibiting aromatase activity. Polyphenol-rich fractions derived from fermented juice, aqueous pericarp extract and cold-pressed or supercritical CO_2-extracted seed oil of pomegranate (Wonderful cultivar) have been reported to inhibit aromatase and 17-beta-hydroxysteroid dehydrogenase type 1 (a key determinant of the increase in estradiol/estrone ratio) activities [92]. The same authors found that the polyphenol-rich fractions from fermented juice and pericarp inhibited the viability of MCF-7 estrogen-dependent tumor cells to a higher extent as compared to estrogen-independent MB-MDA-231 cells; interestingly, normal human breast epithelial cells (MCF-10A) were far less sensitive to the inhibitory effect of polyphenol-rich fractions. Among some other fruits, the ripened fruit and seeds of *Syzygium cumini* (commonly known as *jamun* in Indian subcontinent culture) and *Eugenia jambolana* have also been reported as good reservoir of ellagic acid/ET which, in addition to anthocyanins, can exhibit anti-proliferative properties against various cancer cells [93]. Accordingly, and in strict analogy with the study by Kim *et al.*, [92], *Jamun* fruit extracts have been shown to inhibit over-expressing aromatase and estrogen-dependent MCF-7aro cell proliferation (IC50 27 µg/mL) more effectively as compared to estrogen receptor-negative MDA-MB-231 (IC50 40 µg/mL) breast cancer cells [94]. Pro-apoptotic effects were observed (200 µg/mL) against both MCF-7aro and MDA-MB-231 breast cancer cells, but not toward the normal MCF-10A breast cells.

Upregulation of the phosphoinositide-3 kinase (PI3K)/Akt signaling pathway is a common feature in most human cancers, including breast cancer. Targeting the PI3K pathway with small molecule inhibitors has been studied for therapeutic purposes, and inhibitors such as GDC-0941 or GDC-0980 have entered preclinical trials [95].

Cistaceae family—rock rose family—has been traditionally used in Mediterranean cultures since ancient times. Aqueous extracts recovered from the leaves of *C. populifolius*, which contain high amounts of punicalagin and other ET, have been shown to be cytotoxic against HER 2-dependent (MCF 7/HER2) and -independent (JIMT-1) human breast cancer cells [96]. Since JIMT-1 cells are representative of trastuzumab-resistant cells, *C. populifolius* extracts may be important in the treatment of breast tumors insensitive to this targeted drug.

Finally, oenothein B has proven to exert *in vitro* inhibitory properties against mammary ascites tumors (MM2) cells and Meth-2 solid tumors by releasing interlukin-1 and interlukin-1β-like cytokines [97].

5.4. Oral, Esophageal, and Gastric Cancers

Enzinger and Mayer [98] in their report published in the New England Journal of Medicine indicated esophageal cancer as the deadliest and least-studied type of cancer, with relatively small advancements in diagnosis and treatment over a three decades period. Among other etiological factors of esophageal cancer, inhalation of cigarette smoke is the most obnoxious one in exposing esophageal mucosa to potential carcinogens (*i.e.*, nitrosamines) [99]. Fruits, particularly berries, are a good source of antioxidant including vitamins, anthocyanins, ET, and a wide range of phenolic acids [100]. Consumption of fruits and vegetables has been linked with lower risks of gastrointestinal tract cancer development. This is one of the reasons that prompted researchers to exploit the nutraceutical potential of berries and their biomolecules as chemopreventive food and dietary supplements [101].

As demonstrated by Yoshida *et al.*, [23], high molecular weight oligomeric ET (eucarpanins and elaeagnatins) and macrocyclic dimers including camelliin B, oentothein B, and woodfordin C have cytotoxic properties and induce apoptosis through a pro-oxidant mechanism in tumor cells of oral squamous cell carcinoma (HSC-2, HSG) to a higher extent as compared to normal fibroblasts. These ET are contained in high amounts in flowering plants of *Myrtaceae* and *Elaeagnaceae* family. Black raspberries possess conspicuous quantities of anthocyanins and ET that make them rational candidates for a preventive and therapeutic approach against certain GI tract cancers [102]. Previous studies by Mandal and Stoner [103] and Daniel and Stoner [104] demonstrated that ellagic acid (4 g/kg b.w.) significantly decreased (~60%) the number of *N*-nitrosomethylbenzylamine (NBMA)-induced esophageal tumors in rats. Latter work by Stoner and Morse [105] confirmed the potent anti-tumorigenic property of ellagic acid in rats exposed to NMBA and tobacco nitrosamines through the inhibition of cytochrome P450, which is responsible for the metabolic activation of these carcinogens. Another study by Stoner *et al.*, [100] showed that a lyophilized mix of berries (black raspberries, blackberries, and strawberries) inhibits tumor initiation and progression via downregulation of COX-2 and inducible nitric oxide synthase, events leading to reduced prostaglandin production and nitrate/nitrite levels in the esophagus, respectively.

In a more recent study [106], NBMA-treated rats fed 5%–10% freeze-dried black raspberries showed fewer hyperplastic and dysplastic esophageal lesions, reduced tumor incidence (~54%), multiplicity (~62%), and proliferation as compared to NBMA control rats; more interestingly, it was shown that black raspberries modulate the expression of a panel of genes and proteins involved in the late stages of NMBA-induced rat esophageal tumorigenesis, such as genes involved in carbohydrate and lipid metabolism, cell proliferation and death, inflammation, and proteins involved in cell-cell adhesion, cell proliferation, apoptosis, inflammation, angiogenesis, and both COX and lipoxygenase pathways of arachidonic acid metabolism.

However, the question of which is the relative contribution of ET and anthocyanins to the above chemopreventive activity of berries in esophageal cancer is still open. Indeed, a study by Wang *et al.*, [107] reported that different berries suppress NMBA-induced tumorigenesis irrespective of their ET and anthocyanin content. This finding suggests that also other components of the active preparations of berries, such as lignans and fibers, contribute to the whole chemopreventive capacity, which does not necessarily coincide with the simple sum of the intrinsic activity of each active constituent, but rather depends on positive (or negative) interactions occurring at specific proportions.

Gemin A and B, two ET from *Geum japonicum* Thunb., were found to exert mild cytotoxic effects on human BGC-823 gastric cancer cells [108].

As to oral cancer, Zhang *et al.*, reported that strawberry crude extracts or their isolated components including ellagic acid were toxic toward human oral CAL-27 and KB tumor cells [109]; ellagic acid alone (50–200 μM) exhibited selective cytotoxicity against HSC-2 oral carcinoma cells [110].

Lyophilized strawberries (LS), which carry 42.9% ET and their derivatives and 48.8% anthocyanins, have been referred as an effective option to prevent oral carcinogenesis: indeed a diet containing 5% LS reduced the number of 7,12-dimethylbenz[a]anthracene (DMBA)-induced cheek pouch tumors in hamsters inhibiting Ras/Raf/ERK-dependent cell proliferation, VEGF-dependent angiogenesis, 5-LOX/LTB4 pathway, and prevented oxidative damage [111]; LS was also found to modulate the genetic signature related to DMBA-induced tumor development, such as $p13^{Arf}$, p16, p53, and Bcl-2 [112].

In the same experimental model of hamster buccal pouch carcinoma, it was demonstrated that dietary supplementation of ellagic acid (up to 0.4%) modulated the expression profiles of 37 genes involved in DMBA-induced oral carcinogenesis [113], blocked the development of carcinomas by suppression of Wnt/β-catenin signaling associated with the inactivation of NF-κB and modulation of key components of the mitochondrial apoptotic network [114], and prevented angiogenesis by abrogating hypoxia-driven PI3K/Akt/mTOR, MAPK, and VEGF/VEGFR2 signaling pathways. These effects were mediated by the suppression of histone deacetylase 6 and HIF-1α responses [115].

By virtue of these properties, LS and its major component ellagic acid are considered among the most important and attractive nutraceutical tools for the prevention of oral cancer [116].

5.5. Liver Cancer

Primary liver cancer is, globally, the sixth most frequent cancer, and the second leading cause of cancer death, with a 17% five year survival rate; the leading cause of liver cancer is cirrhosis due to either hepatitis B, hepatitis C, or alcohol [117].

PGG, a major component of *Paeonia suffruticosa* ANDREWS and from *Rhus chinensis* Mill, was found to exhibit *in vitro* antiproliferative activity on human SK-HEP-1 hepatocellular carcinoma cells [118]. The growth-inhibitory effect was related to the ability to cause a G0/G1-phase arrest and to suppress the activation of NF-κB, likely via an IκB-mediated mechanism. PGG was also shown to induce atypical senescence-like S-phase arrest in HepG2 and Huh-7 human hepatocarcinoma cells at sub-lethal doses, increased senescence-associated β-galactosidase activity, and loss of proliferative capacity, through a mechanism involving intracellular generation of oxygen free radicals [119]. No evidence of necrosis or apoptosis was noticed in this study. Interestingly, a more recent report from the same group showed that autophagy was involved in the PGG-induced senescence-like growth arrest, and that activation of MAPK8/9/10 (mitogen-activated protein kinase 8/9/10/c-Jun N-terminal kinases) was an essential upstream signal for autophagy to occur [120]; interestingly, these *in vitro* results were also validated *in vivo* in a xenograft mouse model of human HepG2 liver cancer.

Intraperitoneal administration of corilagin from *Phyllanthus urinaria* was found to significantly reduce the *in vivo* growth of xenografted Hep3B hepatocellular carcinoma cells in athymic nude mice with no adverse effects on liver [121]. Corilagin inhibited the growth of normal or tumor hepatic cells with remarkably different IC50s: indeed the values for normal Chang-liver cells *vs.* the hepatocarcinoma cell lines Bel7402 and SMMC7721 were 131.4 *vs.* 24.5 and 23.4 μM, respectively [122]. The antiproliferative effect in SMMC7721 cells was causally associated with arrest at the G2/M phase by the activation of phospho-p53-p21$^{(Cip1)}$-cdc2/cyclin. Furthermore, a 47.3% growth inhibition was recorded in hepatocarcinoma MHCC97-H cells xenografted in Balb/c mice intraperitoneally treated with 30 mg/kg b.w. corilagin for five weeks.

In a parallel, but different direction, corilagin was found to enhance the cytotoxicity of the reference antitumor drugs cisplatin and doxorubicin on Hep3B hepatoma cells at nutritionally-attainable concentrations [67]. The association of corilagin with low dosages of standard anticancer drugs such as cisplatin or doxorubicin could increment their anticancer effect, enhance their cytotoxic activity toward multi-drug resistant cells, and reduce their toxicity.

Thonningianin A from *Thonningia sanguinea* inhibited the proliferation of HepG-2 human hepatocellular carcinoma cells [123]. Thonningianin A induced caspase-dependent apoptotic cell death, accompanied by an increase in the sub-G1 cell population and DNA fragmentation. Several mechanisms contributing to the antitumor effects were identified: thonningianin A disrupted the mitochondrial membrane potential promoting an increased generation of reactive oxygen species, downregulated the Bcl-xL mRNA expression, induced cell-cycle arrest by changing the cyclin D1 and CDK4 mRNA expression levels. Furthermore, thonningianin A significantly downregulated the NF-κB cell survival pathway concomitantly with the upregulation of the expression level of phosphorylated P38 and downregulation of the expression level of phosphorylated ERK.

5.6. Cervical Cancer

Cervical cancer has long remained a leading cause of malignancies-related death in women from United States of America. However, the number of cervical cancer patients and associated death toll has significantly decreased since last few decades, probably due to the regular Human Papilloma Virus (HPV) screening [124]. Apart from other risk factors, strong association exists between cervical cancers and HPV infection, and HPVs are indicated as central etiological factor in incidents of cervical cancer, globally [125]. Ramasamy *et al.*, [126] found that *Phyllanthus watsonii* extract induced apoptosis in HPV-transformed CaSki epidermoid cervical carcinoma cells, and attributed to the high ellagic content its cytotoxic effect. Raspberry extracts naturally enriched with ET inhibit proliferation of cervical cancer cells (HeLa) in a dose-dependent manner [127]. The study further reported the bound ET-enriched fraction of raspberry extracts as more effective (IC50 = 13 μg/mL) than the unbound anthocyanin-enriched fraction (IC50 = 67 μg/mL).

Hydrolysable tannins improve dysfunctional gap junctions communication, which are involved in carcinogenesis. Tellimagrandin I and chebulinic acid restore dysfunctional gap junctions in HeLa cells. *In vitro* exposure of HeLa cells to tellimagrandin I inhibits their proliferation as well as their substrate-independent growth [128].

Camelliin B, the hydrolysable tannin isolated from a non-edible plant (*i.e.*, *Gordonia axillaris* or fried eggplant), is another example of phytochemical useful for cervical cancer treatment. Camelliin B isolated from *G. axillaris* inhibited the growth of HeLa cells with an IC50 of 46.3 μg/mL as compared to the IC50 of 108.0 μg/mL observed in normal cervical fibroblasts [129]. The study showed that camelliin B induces chromatin condensation, a hallmark of apoptosis. Furthermore, camelliin B also exhibited DNA fragmentation properties and inhibited the DNA repair-associated enzyme poly (ADP-ribose) polymerase in HeLa cells. Walnut extracts rich in tellimagrandin I and II induce cytotoxic effects in human HeLa cancer cells by reducing mitochondrial respiration and promoting apoptosis [130]. Ellagic acid was shown to induce G1 arrest via induction of p21 and apoptosis in CaSki human cervix carcinoma cells [59].

The elevated risk of cervical cancer in cigarette smokers is thought to depend on the increased mutations in cervical cells caused by the persistence of smoke habit-associated DNA damage in the presence of HPV infection. Importantly, ellagic acid significantly attenuates cigarette smoke-induced DNA damage in HPV16-transformed human ECT1/E6 E7 ectocervical cells [131], an effect which is likely to derive from ellagic acid antioxidant and free-radical scavenging activity and that further support its chemopreventive potential.

5.7. Lung Cancer

Lung cancer is the most prevalent cancer worldwide [68]. The prognosis of lung cancer patients is still poor, and while it is not the most frequently diagnosed cancer in the United States, it is by far the leading cause of cancer-related deaths in the US and also worldwide. Therefore, advances in the treatment of lung cancer are urgently needed.

Although the relative importance of its major constituents, ET and anthocyanins, was not addressed, pomegranate extracts have been found to exert antiproliferative and chemopreventive activities against lung cancer *in vitro* and in animal models [132,133]. Other reports suggest a specific and important role for ET in the pomegranate extract activity against this type of malignancy, in both *in vitro* and animal experimental settings. In a study focusing on purified ellagic acid and punicalagin, Zahin *et al.*, [134] demonstrated that these two compounds were antimutagenic, prevented the formation of benzo[a]pyrene-induced DNA adducts, and were antiproliferative in non-small cell lung cancer A549 and H1299 lung cancer cells. It is worth noting that punicalagin, using the same toxicity tetrazolium assay, had been shown to be far less antiproliferative toward the same A549 cell line [135] as compared to the data reported by Zahin *et al*, [134]. This apparent discrepancy, which points to the importance of standardizing the experimental settings in this kind of studies, is likely to depend on the post-treatment incubation times before determining cell viability: in the first study, cell viability was determined at 24 h [121], while in the second one at 48 h [122], a time which allows a more accurate estimate of the growth inhibitory activity. Kuo *et al.*, [136] found that the ET casuarinin from the bark of *Terminalia arjuna* induced apoptosis in human breast adenocarcinoma MCF-7 cells and in A549 cells by blocking cell-cycle progression in the G0/G1 phase.

Similarly, *jamun* (*Syzygium cumini* L.) seeds and pulp hydrolyzed extracts have been reported to exert antiproliferative activity in A549 cells, which has been associated with the presence of ellagic acid [93].

5.8. Skin Cancer

Prolonged exposure of skin to UV radiation is causally linked to several pathological conditions, including photo-aging and photocarcinogenesis. UV damage is partly attributable to increased skin reactive oxygen species generation. Pomegranate fruit extract, which contains very high amounts of ET, has been shown to exert a significant protective effect against UV rays insult and pathological consequences. Orally-administered pomegranate extract containing 90% ellagic acid, by virtue of its antioxidant activity, has been shown to inhibit skin pigmentation induced by exposure to UV radiation in brown guinea pigs [137]; under the same conditions, the extract decreased melanocyte proliferation and melanin synthesis via inhibition of tyrosinase activity to a degree comparable to that of arbutin, an established tyrosinase inhibitor.

Several studies have confirmed the ability of standardized pomegranate extract and pomegranate ET (500–10,000 mg/L) to inhibit free radical generation in UVA- and UVB-irradiated human skin, thus protecting it from DNA fragmentation, skin burns, and pigmentation, and finally decreasing the risk of malignant transformation [4]. Various mechanisms involved include reduction of DNA damage, prevention of UVB-caused matrix metalloproteinases induction, inhibition of matrix metalloproteinases 2 and 9 activity, and decrease in UVB-induced c-Fos protein expression and c-Jun phosphorylation [138].

Animal studies further confirmed the chemopreventive and anticancer activity of ET-rich pomegranate extract: in a UVB initiation-promotion protocol, SKH-1 hairless mice receiving oral pomegranate extract supplementation showed reduced tumor incidence, prolonged latency periods of tumor appearance, and lower tumor body burden compared to that of unsupplemented UVB-irradiated control animals [139].

6. Risks and Safe Consumption Levels

In contrast with the widely accepted notion that ET, similarly to other phytochemicals, are health-promoting, chemopreventive, and therapeutically-valuable compounds, data emerged from some studies raised the question of the safety of their consumption [140]. In general, tannins may be toxic to cells and tissues because of their protein precipitation, enzymes inhibition, and mineral binding properties [140,141]. Furthermore, it was reported that pomegranate hydroalcoholic extract exerts mutagenic, genotoxic and clastogenic effects in a panel of *in vitro* and *in vivo* assays [142]. In Chinese hamster B14 cells, ellagic acid and gallic acid caused the production of DNA single-strand breaks with no relation to the concentration used, cytotoxic effects and increased lipid bilayer fluidity, an event which the authors suggested as contributing to DNA single-strand breakage [143]. However, these results are controversial and contradicted by studies demonstrating the lack of mutagenicity of ellagic acid in similar experimental settings [144] and by the hundreds of reports on the DNA protective activity of polyphenols, including ET, against established genotoxic agents.

A study conducted by Filippich *et al.*, [145] linked the generation of lesions on mice liver, early and severe liver necrosis, to punicalagin. However, an update on punicalagin risk assessment revealed neither hepatotoxic nor nephrotoxic effects following sub-chronic oral exposure (6% daily) to Sprague–Dawley rats [146].

ET have been reported to act as α-glucosidase inhibitors and, thus, proposed as adjunctive agents in type-2 diabetes management [147]: a caveat has been associated with this property since the dietary intake of any α-glucosidase inhibitor in normal circumstances might generate risks of carbohydrate malabsorption, gastrointestinal discomfort, flatulence, and diarrhea, such as for acarbose [148,149]. However, to the best of our knowledge, there is no report of such side effects causally linked to ingestion of ET-rich food and fruits.

ET, alongside the condensed tannins, could be considered as antinutritional in animal diets due to their ability of interacting with protein and inhibiting certain enzymes. Antinutritional effects have been reported in animal models, where diet carrying tannins at dosages higher than 10 g/kg b.w. affected animal growth and digestive capacity [150]. However, levels ⩾10 g/kg b.w. are unlikely to be attained using standard nutritional regimens; furthermore, a study conducted for risk assessment of chestnut hydrolysable tannins included in lamb diet revealed the lack of any toxic response in terms of weight gain, protein conversion efficiency, and histopathological features [151].

To date, incomplete information is available on toxicity and risk assessment of individual ET. However, the no observed effect levels (NOEL) and no observed adverse effect levels (NOAEL) as determined in some reports are unlikely to portray dietary consumption-associated toxicity. For example, a 90-day sub-chronic toxicity study performed in F344 rats showed that ellagic acid NOEL was 3011 mg/kg b.w./day for males and the NOAEL and NOEL in females were 3254 mg/kg b.w./day and ⩽778 mg/kg b.w./day, respectively, and there were no obvious histopathological changes in any of the groups [152]. A 90-day sub-chronic study showed that the LD50 of a standardized pomegranate fruit extract containing 30% punicalagin in Wistar rats was >5 g/kg b.w., with no visible sign of toxicity in terms of feed consumption, weight gain, ophthalmic, and pathological evaluation [153].

Dietary intake of ET varies among cultures, communities and region as has been evidently documented in studies from different countries [6,154]. A global report on the dietary consumption of phytonutrients reveals that peoples from Western Europe have maximum ellagic acid consumption trends in both genders (7.6 mg/day in males and 7.9 mg/day in females). Berries account approximately for 90% of the daily ellagic acid intake [154]. A few reports on the nutritional habits of

German and Finnish communities indicate that consumption of berries provides up to 5 mg and 12 mg ET per day, respectively [6,155,156]. Correlating ET consumption trends from various dietary sources with the so far identified NOEL or NOAEL for these biomolecules undoubtedly indicate that ET pose negligible threats to the safety and health security of the consumers, consolidating the notion that ET, either in individual or composite form, can potentially be exploited as health-promoting and potential chemopreventive phytonutrients.

As a final consideration, it could be speculated that an increasing use of ET as anticancer agents could pave the way to the adoption of administration routes different from oral one, such as the intravenous administration: such a route, however, would need to be characterized from the toxicological point of view since this kind of data is still lacking.

7. Concluding Remarks

The increasing awareness and knowledge of the capacity of plant-derived compounds to modify cell transformation and cancer cell growth suggest that they could serve as new tools for either preventive and therapeutic interventions. Today, ET are recognized as a class of phytochemicals characterized by a strong potential for development as chemopreventive, and possibly as therapeutic, agents against various human cancers. This could have a direct clinical and translational relevance to cancer patients if consumption of ET-rich fruits and vegetables will unequivocally prove to contrast the process of carcinogenesis and tumor growth, with positive outcomes in terms of survival and quality of life of the patient. To this end, future research should be addressed to define the actual clinical potential of ET through specific studies such as the determination of the systemic bioavailability from either food sources or concentrated formulations, the optimal period of administration and dosing, the toxicity and side effects (if any), the anticancer activity. The effects of single ET and of rational combinations of different ET should also be addressed. A multidisciplinary and coordinated approach will be needed and will involve basic research investigations, epidemiological and preclinical studies including the effect of combining ET with conventional antineoplastic drugs.

Author Contributions: All the authors contributed to the writing and revision of this review article.

Conflicts of Interest: The authors declare no conflict of interest.

References

1. Siegel, R.; Ward, E.; Brawley, O.; Jemal, A. The impact of eliminating socioeconomic and racial disparities on premature cancer deaths. *CA Cancer J. Clin.* **2011**, *61*, 212–236.
2. Durko, L.; Malecka-Panas, E. Lifestyle modifications and colorectal cancer. *Curr. Colorectal. Cancer Rep.* **2014**, *10*, 45–54. [CrossRef] [PubMed]
3. Quideau, S.; Deffieux, D.; Douat-Casassus, C.; Pouysegu, L. Plant polyphenols: Chemical properties, biological activities, and synthesis. *Angew. Chem. Int. Ed. Engl.* **2011**, *50*, 586–621. [CrossRef] [PubMed]
4. Ismail, T.; Sestili, P.; Akhtar, S. Pomegranate peel and fruit extracts: A review of potential anti-inflammatory and anti-infective effects. *J. Ethnopharmacol.* **2012**, *143*, 397–405. [CrossRef] [PubMed]
5. Landete, J. Ellagitannins, ellagic acid and their derived metabolites: A review about source, metabolism, functions and health. *Food Res. Int.* **2011**, *44*, 1150–1160. [CrossRef]
6. Koponen, J.M.; Happonen, A.M.; Mattila, P.H.; Torronen, A.R. Contents of anthocyanins and ellagitannins in selected foods consumed in Finland. *J. Agric. Food Chem.* **2007**, *55*, 1612–1619. [CrossRef] [PubMed]
7. Mullen, W.; McGinn, J.; Lean, M.E.; MacLean, M.R.; Gardner, P.; Duthie, G.G.; Yokota, T.; Crozier, A. Ellagitannins, flavonoids, and other phenolics in red raspberries and their contribution to antioxidant capacity and vasorelaxation properties. *J. Agric. Food Chem.* **2002**, *50*, 5191–5196. [CrossRef] [PubMed]
8. Mullen, W.; Stewart, A.J.; Lean, M.E.; Gardner, P.; Duthie, G.G.; Crozier, A. Effect of freezing and storage on the phenolics, ellagitannins, flavonoids, and antioxidant capacity of red raspberries. *J. Agric. Food Chem.* **2002**, *50*, 5197–5201. [CrossRef] [PubMed]

9.   Gancel, A.L.; Feneuil, A.; Acosta, O.; Perez, A.M.; Vaillant, F. Impact of industrial processing and storage on major polyphenols and the antioxidant capacity of tropical highland blackberry (*Rubus adenotrichus*). *Food Res. Int.* **2011**, *44*, 2243–2251. [CrossRef]

10. Hager, T.J.; Howard, L.R.; Liyanage, R.; Lay, J.O.; Prior, R.L. Ellagitannin composition of blackberry as determined by HPLC-ESI-MS and MALDI-TOF-MS. *J. Agric. Food Chem.* **2008**, *56*, 661–669. [CrossRef] [PubMed]

11. Mertz, C.; Cheynier, V.; Gunata, Z.; Brat, P. Analysis of phenolic compounds in two blackberry species (*Rubus glaucus* and *Rubus adenotrichus*) by high-performance liquid chromatography with diode array detection and electrospray ion trap mass spectrometry. *J. Agric. Food Chem.* **2007**, *55*, 8616–8624. [CrossRef] [PubMed]

12. Lansky, E.P.; Newman, R.A. *Punica granatum* (pomegranate) and its potential for prevention and treatment of inflammation and cancer. *J. Ethnopharmacol.* **2007**, *109*, 177–206. [CrossRef] [PubMed]

13. Beekwilder, J.; Jonker, H.; Meesters, P.; Hall, R.D.; van der Meer, I.M.; de Vos, C.H.R. Antioxidants in raspberry: On-line analysis links antioxidant activity to a diversity of individual metabolites. *J. Agric. Food Chem.* **2005**, *53*, 3313–3320. [CrossRef] [PubMed]

14. Lee, J.H.; Johnson, J.V.; Talcott, S.T. Identification of ellagic acid conjugates and other polyphenolics in muscadine grapes by HPLC-ESI-MS. *J. Agric. Food Chem.* **2005**, *53*, 6003–6010. [CrossRef] [PubMed]

15. Maatta-Riihinen, K.R.; Kamal-Eldin, A.; Torronen, A.R. Identification and quantification of phenolic compounds in berries of *Fragaria* and *Rubus* species (family Rosaceae). *J. Agric. Food Chem.* **2004**, *52*, 6178–6187. [CrossRef] [PubMed]

16. Mullen, W.; Yokota, T.; Lean, M.E.; Crozier, A. Analysis of ellagitannins and conjugates of ellagic acid and quercetin in raspberry fruits by LC-MS^n. *Phytochemistry* **2003**, *64*, 617–624. [CrossRef]

17. Rangkadilok, N.; Worasuttayangkurn, L.; Bennett, R.N.; Satayavivad, J. Identification and quantification of polyphenolic compounds in longan (*Euphoria longana* Lam.) fruit. *J. Agric. Food Chem.* **2005**, *53*, 1387–1392. [CrossRef] [PubMed]

18. Seeram, N.P.; Lee, R.; Heber, D. Bioavailability of ellagic acid in human plasma after consumption of ellagitannins from pomegranate (*Punica granatum* L.) juice. *Clin. Chim. Acta* **2004**, *348*, 63–68. [CrossRef] [PubMed]

19. Abe, L.T.; Lajolo, F.M.; Genovese, M.I. Comparison of phenol content and antioxidant capacity of nuts. *Food Sci. Technol. (Camp.)* **2010**, *30*, 254–259. [CrossRef]

20. Kaume, L.; Howard, L.R.; Devareddy, L. The blackberry fruit: A review on its composition and chemistry, metabolism and bioavailability, and health benefits. *J. Agric. Food Chem.* **2012**, *60*, 5716–5727. [CrossRef] [PubMed]

21. Shi, N.; Clinton, S.K.; Liu, Z.; Wang, Y.; Riedl, K.M.; Schwartz, S.J.; Zhang, X.; Pan, Z.; Chen, T. Strawberry phytochemicals inhibit azoxymethane/dextran sodium sulfate-induced colorectal carcinogenesis in Crj: CD-1 mice. *Nutrients* **2015**, *7*, 1696–1715. [CrossRef] [PubMed]

22. Yoshida, T.; Ito, H.; Hatano, T.; Kurata, M.; Nakanishi, T.; Inada, A.; Murata, H.; Inatomi, Y.; Matsuura, N.; Ono, K.; *et al.* New hydrolyzable tannins, Shephagenins A and B, from shepherdia argentea as HIV-1 reverse transcriptase inhibitors. *Chem. Pharm. Bull.* **1996**, *44*, 1436–1439. [CrossRef] [PubMed]

23. Yoshida, T.; Hatano, T.; Ito, H. Chemistry and function of vegetable polyphenols with high molecular weights. *BioFactors* **2000**, *13*, 121–125. [CrossRef] [PubMed]

24. Harborne, J.B. *Plant Phenolics*; Academic Press Ltd.: Chicago, IL, USA, 1983; Volume 1.

25. Haslam, E. *Plant Polyphenols: Vegetable Tannins Revisited*; CUP Archive: Cambridge, UK, 1989.

26. Haslam, E.; Cai, Y. Plant polyphenols (vegetable tannins): Gallic acid metabolism. *Nat. Prod. Rep.* **1994**, *11*, 41–66. [CrossRef] [PubMed]

27. Khanbabaee, K.; van Ree, T. Tannins: Classification and definition. *Nat. Prod. Rep.* **2001**, *18*, 641–649.

28. Salminen, J.P.; Karonen, M. Chemical ecology of tannins and other phenolics: We need a change in approach. *Funct. Ecol.* **2011**, *25*, 325–338. [CrossRef]

29. Gross, G.G. From lignins to tannins: Forty years of enzyme studies on the biosynthesis of phenolic compounds. *Phytochemistry* **2008**, *69*, 3018–3031. [CrossRef] [PubMed]

30. Larrosa, M.; García-Conesa, M.T.; Espín, J.C.; Tomás-Barberán, F.A. *Bioavailability and Metabolism of Ellagic Acid and Ellagitannins*; CRC Press: Boca Raton, FL, USA, 2012.

31. Larrosa, M.; Tomas-Barberan, F.A.; Espin, J.C. The dietary hydrolysable tannin punicalagin releases ellagic acid that induces apoptosis in human colon adenocarcinoma Caco-2 cells by using the mitochondrial pathway. *J. Nutr. Biochem.* **2006**, *17*, 611–625. [CrossRef] [PubMed]

32. Kilkowski, W.J.; Gross, G.G. Color reaction of hydrolyzable tannins with bradford reagent, coomassie brilliant blue. *Phytochemistry* **1999**, *51*, 363–366. [CrossRef]

33. Barbehenn, R.V.; Jones, C.P.; Hagerman, A.E.; Karonen, M.; Salminen, J.P. Ellagitannins have greater oxidative activities than condensed tannins and galloyl glucoses at high pH: Potential impact on caterpillars. *J. Chem. Ecol.* **2006**, *32*, 2253–2267. [CrossRef] [PubMed]

34. Barbehenn, R.V.; Jones, C.P.; Karonen, M.; Salminen, J.P. Tannin composition affects the oxidative activities of tree leaves. *J. Chem. Ecol.* **2006**, *32*, 2235–2251. [CrossRef] [PubMed]

35. Quideau, S. *Chemistry and Biology of Ellagitannins: An Underestimated Class of Bioactive Plant Polyphenols*; World Scientific: Hackensack, NJ, USA, 2009.

36. Okuda, T.; Yoshida, T.; Hatano, T. Hydrolyzable tannins and related polyphenols. In *Fortschritte der Chemie Organischer Naturstoffe/Progress in the Chemistry of Organic Natural Products*; Springer: Vienna, Austria, 1995; pp. 1–117.

37. Okuda, T.; Yoshida, T.; Hatano, T. Correlation of oxidative transformations of hydrolyzable tannins and plant evolution. *Phytochemistry* **2000**, *55*, 513–529. [CrossRef]

38. Okuda, T.; Yoshida, T.; Ashida, M.; Yazaki, K. Tannis of *Casuarina* and *Stachyurus* species. Part 1. Structures of pendunculagin, casuarictin, strictinin, casuarinin, casuariin, and stachyurin. *J. Chem. Soc. Perkin Trans.* **1983**, *1*, 1765–1772. [CrossRef]

39. Okuda, T.; Yoshida, T.; Kuwahara, M.; Memon, M.U.; Shingu, T. Agrimoniin and potentillin, an ellagitannin dimer and monomer having an α-glucose core. *J. Chem. Soc. Chem. Commun.* **1982**, 163–164. [CrossRef]

40. Okuda, T.; Yoshida, T.; Kuwahara, M.; Memon, M.U.; Shingu, T. Tannins of rosaceous medicinal plants. I. Structures of potentillin, agrimonic acids A and B, and agrimoniin, a dimeric ellagitannin. *Chem. Pharm. Bull.* **1984**, *32*, 2165–2173. [CrossRef]

41. Lipińska, L.; Klewicka, E.; Sójka, M. Structure, occurrence and biological activity of ellagitannins: A general review. *Acta Sci. Pol. Technol. Aliment.* **2014**, *13*, 289–299. [CrossRef] [PubMed]

42. Kaneshima, T.; Myoda, T.; Nakata, M.; Fujimori, T.; Toeda, K.; Nishizawa, M. Antioxidant activity of *C*-glycosidic ellagitannins from the seeds and peel of camu-camu (*Myrciaria dubia*). *LWT Food Sci. Technol.* **2016**, *69*, 76–81. [CrossRef]

43. Tanaka, T.; Ueda, N.; Shinohara, H.; Nonaka, G.-I.; Kouno, I. Four new *C*-glycosidic ellagitannins, castacrenins DG, from Japanese chestnut wood (castanea crenata SIEB. Et ZUCC.). *Chem. Pharm. Bull.* **1997**, *45*, 1751–1755. [CrossRef]

44. Omar, M.; Matsuo, Y.; Maeda, H.; Saito, Y.; Tanaka, T. New metabolites of *C*-glycosidic ellagitannin from Japanese oak sapwood. *Org Lett.* **2014**, *16*, 1378–1381. [CrossRef] [PubMed]

45. Jiang, Z.-H.; Tanaka, T.; Kouno, I. Three novel *C*-glycosidic ellagitannins, Rhoipteleanins H, I, and J, from Rhoiptelea c hiliantha. *J. Nat. Prod.* **1999**, *62*, 425–429. [CrossRef] [PubMed]

46. Quideau, S.; Jourdes, M.; Lefeuvre, D.; Montaudon, D.; Saucier, C.; Glories, Y.; Pardon, P.; Pourquier, P. The chemistry of wine polyphenolic *C*-glycosidic ellagitannins targeting human topoisomerase II. *Chemistry* **2005**, *11*, 6503–6513. [CrossRef] [PubMed]

47. Clifford, M.N.; Scalbert, A. Ellagitannins—Nature, occurrence and dietary burden. *J. Sci. Food Agric.* **2000**, *80*, 1118–1125. [CrossRef]

48. Garcia-Munoz, C.; Vaillant, F. Metabolic fate of ellagitannins: Implications for health, and research perspectives for innovative functional foods. *Crit. Rev. Food Sci. Nutr.* **2014**, *54*, 1584–1598. [CrossRef] [PubMed]

49. Bialonska, D.; Kasimsetty, S.G.; Khan, S.I.; Ferreira, D. Urolithins, intestinal microbial metabolites of pomegranate ellagitannins, exhibit potent antioxidant activity in a cell-based assay. *J. Agric. Food Chem.* **2009**, *57*, 10181–10186. [CrossRef] [PubMed]

50. González-Barrio, R.O.; Borges, G.; Mullen, W.; Crozier, A. Bioavailability of anthocyanins and ellagitannins following consumption of raspberries by healthy humans and subjects with an ileostomy. *J. Agric. Food Chem.* **2010**, *58*, 3933–3939. [CrossRef] [PubMed]

51. Garcia-Munoz, C.; Hernàndez, L.; Pèrez, A.; Vaillant, F. Diversity of urinary excretion patterns of main ellagitannins' colonic metabolites after ingestion of tropical highland blackberry (*Rubus adenotrichus*) juice. *Food Res. Int.* **2014**, *55*, 161–169. [CrossRef]

52. Seeram, N.P.; Lee, R.; Scheuller, H.S.; Heber, D. Identification of phenolic compounds in strawberries by liquid chromatography electrospray ionization mass spectroscopy. *Food Chem.* **2006**, *97*, 1–11. [CrossRef]

53. Cerdá, B.; Espín, J.C.; Parra, S.; Martínez, P.; Tomás-Barberán, F.A. The potent *in vitro* antioxidant ellagitannins from pomegranate juice are metabolised into bioavailable but poor antioxidant hydroxy-6*H*-dibenzopyran-6-one derivatives by the colonic microflora of healthy humans. *Eur. J. Nutr.* **2004**, *43*, 205–220. [CrossRef] [PubMed]

54. Larrosa, M.; Garcia-Conesa, M.T.; Espin, J.C.; Tomas-Barberan, F.A. Ellagitannins, ellagic acid and vascular health. *Mol. Asp. Med.* **2010**, *31*, 513–539. [CrossRef] [PubMed]

55. Heber, D. Multitargeted therapy of cancer by ellagitannins. *Cancer Lett.* **2008**, *269*, 262–268. [CrossRef] [PubMed]

56. Tomás-Barberán, F.A.; García-Villalba, R.; González-Sarrías, A.; Selma, M.V.; Espín, J.C. Ellagic acid metabolism by human gut microbiota: Consistent observation of three urolithin phenotypes in intervention trials, independent of food source, age, and health status. *J. Agric. Food Chem.* **2014**, *62*, 6535–6538. [CrossRef] [PubMed]

57. Nicoli, M.; Anese, M.; Parpinel, M. Influence of processing on the antioxidant properties of fruit and vegetables. *Trends Food Sci. Technol.* **1999**, *10*, 94–100. [CrossRef]

58. Balkwill, F.; Coussens, L.M. Cancer: An inflammatory link. *Nature* **2004**, *431*, 405–406. [CrossRef] [PubMed]

59. Narayanan, B.A.; Geoffroy, O.; Willingham, M.C.; Re, G.G.; Nixon, D.W. p53/p21(WAF1/CIP1) expression and its possible role in G1 arrest and apoptosis in ellagic acid treated cancer cells. *Cancer Lett.* **1999**, *136*, 215–221. [CrossRef]

60. Vanella, L.; di Giacomo, C.; Acquaviva, R.; Barbagallo, I.; Cardile, V.; Kim, D.H.; Abraham, N.G.; Sorrenti, V. Apoptotic markers in a prostate cancer cell line: Effect of ellagic acid. *Oncol. Rep.* **2013**, *30*, 2804–2810. [PubMed]

61. Vicinanza, R.; Zhang, Y.; Henning, S.M.; Heber, D. Pomegranate juice metabolites, ellagic acid and urolithin a, synergistically inhibit androgen-independent prostate cancer cell growth via distinct effects on cell cycle control and apoptosis. *Evid. Based Complement. Altern. Med.* **2013**, *2013*, 1–12. [CrossRef] [PubMed]

62. Chen, H.-S.; Bai, M.-H.; Zhang, T.; Li, G.-D.; Liu, M. Ellagic acid induces cell cycle arrest and apoptosis through TGF-β/Smad3 signaling pathway in human breast cancer MCF-7 cells. *Int. J. Oncol.* **2015**, *46*, 1730–1738. [CrossRef]

63. Wen, X.Y.; Wu, S.Y.; Li, Z.Q.; Liu, Z.Q.; Zhang, J.J.; Wang, G.F.; Jiang, Z.H.; Wu, S.G. Ellagitannin (BJA3121), an anti-proliferative natural polyphenol compound, can regulate the expression of miRNAs in HepG2 cancer cells. *Phytother. Res.* **2009**, *23*, 778–784. [CrossRef] [PubMed]

64. Seeram, N.P.; Adams, L.S.; Henning, S.M.; Niu, Y.; Zhang, Y.; Nair, M.G.; Heber, D. *In vitro* antiproliferative, apoptotic and antioxidant activities of punicalagin, ellagic acid and a total pomegranate tannin extract are enhanced in combination with other polyphenols as found in pomegranate juice. *J. Nutr. Biochem.* **2005**, *16*, 360–367. [CrossRef] [PubMed]

65. Sartippour, M.R.; Seeram, N.P.; Rao, J.Y.; Moro, A.; Harris, D.M.; Henning, S.M.; Firouzi, A.; Rettig, M.B.; Aronson, W.J.; Pantuck, A.J. Ellagitannin-rich pomegranate extract inhibits angiogenesis in prostate cancer *in vitro* and *in vivo*. *Int. J. Oncol.* **2008**, *32*, 475–480. [CrossRef] [PubMed]

66. Lee, S.-J.; Lee, H.-K. Sanguiin H-6 blocks endothelial cell growth through inhibition of VEGF binding to VEGF receptor. *Arch. Pharmacal. Res.* **2005**, *28*, 1270–1274. [CrossRef]

67. Gambari, R.; Hau, D.K.P.; Wong, W.Y.; Chui, C.H. Sensitization of Hep3B hepatoma cells to cisplatin and doxorubicin by corilagin. *Phytotherapy Res.* **2014**, *28*, 781–783. [CrossRef] [PubMed]

68. CDC. 2012 Top Ten Cancers. Available online: https://nccd.cdc.gov/uscs/toptencancers.aspx (accessed on 29 January 2016).

69. Masko, E.M.; Allott, E.H.; Freedland, S.J. The relationship between nutrition and prostate cancer: Is more always better? *Eur. Urol.* **2013**, *63*, 810–820. [CrossRef] [PubMed]

70. Cohen, J.H.; Kristal, A.R.; Stanford, J.L. Fruit and vegetable intakes and prostate cancer risk. *J. Natl. Cancer Inst.* **2000**, *92*, 61–68. [CrossRef] [PubMed]

71. Kolonel, L.N.; Hankin, J.H.; Whittemore, A.S.; Wu, A.H.; Gallagher, R.P.; Wilkens, L.R.; John, E.M.; Howe, G.R.; Dreon, D.M.; West, D.W.; *et al.* Vegetables, fruits, legumes and prostate cancer: A multiethnic case-control study. *Cancer Epidemiol. Biomark. Prev.* **2000**, *9*, 795–804.
72. Seeram, N.P.; Aronson, W.J.; Zhang, Y.; Henning, S.M.; Moro, A.; Lee, R.-P.; Sartippour, M.; Harris, D.M.; Rettig, M.; Suchard, M.A. Pomegranate ellagitannin-derived metabolites inhibit prostate cancer growth and localize to the mouse prostate gland. *J. Agric. Food Chem.* **2007**, *55*, 7732–7737. [CrossRef] [PubMed]
73. Albrecht, M.; Jiang, W.; Kumi-Diaka, J.; Lansky, E.P.; Gommersall, L.M.; Patel, A.; Mansel, R.E.; Neeman, I.; Geldof, A.A.; Campbell, M.J. Pomegranate extracts potently suppress proliferation, xenograft growth, and invasion of human prostate cancer cells. *J. Med. Food* **2004**, *7*, 274–283. [CrossRef] [PubMed]
74. Malik, A.; Afaq, F.; Sarfaraz, S.; Adhami, V.M.; Syed, D.N.; Mukhtar, H. Pomegranate fruit juice for chemoprevention and chemotherapy of prostatesystemic antioxidant propo cancer. *Proc. Natl. Acad. Sci. USA* **2005**, *102*, 14813–14818. [CrossRef] [PubMed]
75. Stolarczyk, M.; Piwowarski, J.P.; Granica, S.; Stefanska, J.; Naruszewicz, M.; Kiss, A.K. Extracts from *Epilobium* sp. Herbs, their components and gut microbiota metabolites of epilobium ellagitannins, urolithins, inhibit hormone-dependent prostate cancer cells-(lNCaP) proliferation and PSA secretion. *Phytother. Res.* **2013**, *27*, 1842–1848. [CrossRef] [PubMed]
76. Stolarczyk, M.; Naruszewicz, M.; Kiss, A.K. Extracts from *Epilobium* sp. Herbs induce apoptosis in human hormone-dependent prostate cancer cells by activating the mitochondrial pathway. *J. Pharm. Pharmacol.* **2013**, *65*, 1044–1054. [CrossRef] [PubMed]
77. Walia, H.; Arora, S. Terminalia chebula—A pharmacognistic account. *J. Med. Plant Res.* **2013**, *7*, 1351–1361.
78. Saleem, A.; Husheem, M.; Harkonen, P.; Pihlaja, K. Inhibition of cancer cell growth by crude extract and the phenolics of *Terminalia chebula* retz. Fruit. *J. Ethnopharmacol.* **2002**, *81*, 327–336. [CrossRef]
79. Calixto, J.B. Twenty-five years of research on medicinal plants in Latin America: A personal view. *J. Ethnopharmacol.* **2005**, *100*, 131–134. [CrossRef] [PubMed]
80. Eberhart, C.E.; Coffey, R.J.; Radhika, A.; Giardiello, F.M.; Ferrenbach, S.; Dubois, R.N. Up-regulation of cyclooxygenase 2 gene expression in human colorectal adenomas and adenocarcinomas. *Gastroenterology* **1994**, *107*, 1183–1188. [PubMed]
81. Fajardo, A.M.; Piazza, G.A. Chemoprevention in gastrointestinal physiology and disease. Anti-inflammatory approaches for colorectal cancer chemoprevention. *Am. J. Physiol. Gastrointest. Liver Physiol.* **2015**, *309*, G59–G70. [CrossRef] [PubMed]
82. Madka, V.; Rao, C.V. Anti-inflammatory phytochemicals for chemoprevention of colon cancer. *Curr. Cancer Drug Targets* **2013**, *13*, 542–557. [CrossRef] [PubMed]
83. Adams, L.S.; Seeram, N.P.; Aggarwal, B.B.; Takada, Y.; Sand, D.; Heber, D. Pomegranate juice, total pomegranate ellagitannins, and punicalagin suppress inflammatory cell signaling in colon cancer cells. *J. Agric. Food Chem.* **2006**, *54*, 980–985. [CrossRef] [PubMed]
84. Kasimsetty, S.G.; Bialonska, D.; Reddy, M.K.; Ma, G.; Khan, S.I.; Ferreira, D. Colon cancer chemopreventive activities of pomegranate ellagitannins and urolithins. *J. Agric. Food Chem.* **2010**, *58*, 2180–2187. [CrossRef] [PubMed]
85. Sharma, M.; Li, L.; Celver, J.; Killian, C.; Kovoor, A.; Seeram, N.P. Effects of fruit ellagitannin extracts, ellagic acid, and their colonic metabolite, urolithin a, on Wnt signaling. *J. Agric. Food Chem.* **2009**, *58*, 3965–3969. [CrossRef] [PubMed]
86. CDC. Breast Cancer Statistics. Available online: http://www.cdc.gov/cancer/breast/statistics/ (accessed on 2 February 2016).
87. Russo, I.H.; Russo, J. Role of hormones in mammary cancer initiation and progression. *J. Mammary Gland Biol. Neoplasia* **1998**, *3*, 49–61. [CrossRef] [PubMed]
88. Gebre-Medhin, M.; Kindblom, L.-G.; Wennbo, H.; Törnell, J.; Meis-Kindblom, J.M. Growth hormone receptor is expressed in human breast cancer. *Am. J. Pathol.* **2001**, *158*, 1217–1222. [CrossRef]
89. Chen, Z.; Gu, K.; Zheng, Y.; Zheng, W.; Lu, W.; Shu, X.O. The use of complementary and alternative medicine among Chinese women with breast cancer. *J. Altern. Complement. Med.* **2008**, *14*, 1049–1055. [CrossRef] [PubMed]
90. Brodie, A.; Sabnis, G.; Jelovac, D. Aromatase and breast cancer. *J. Steroid Biochem. Mol. Biol.* **2006**, *102*, 97–102. [CrossRef] [PubMed]
91. Chen, S. Aromatase and breast cancer. *Front. Biosci.* **1998**, *3*, d922–d933. [CrossRef] [PubMed]

92. Kim, N.D.; Mehta, R.; Yu, W.; Neeman, I.; Livney, T.; Amichay, A.; Poirier, D.; Nicholls, P.; Kirby, A.; Jiang, W. Chemopreventive and adjuvant therapeutic potential of pomegranate (*Punica granatum*) for human breast cancer. *Breast Cancer Res. Treat.* **2002**, *71*, 203–217. [CrossRef] [PubMed]

93. Aqil, F.; Gupta, A.; Munagala, R.; Jeyabalan, J.; Kausar, H.; Sharma, R.J.; Singh, I.P.; Gupta, R.C. Antioxidant and antiproliferative activities of anthocyanin/ellagitannin-enriched extracts from *Syzygium cumini* L. (Jamun, the Indian Blackberry). *Nutr. Cancer* **2012**, *64*, 428–438. [CrossRef] [PubMed]

94. Li, L.; Adams, L.S.; Chen, S.; Killian, C.; Ahmed, A.; Seeram, N.P. *Eugenia jambolana* lam. Berry extract inhibits growth and induces apoptosis of human breast cancer but not non-tumorigenic breast cells. *J. Agric. Food Chem.* **2009**, *57*, 826–831. [CrossRef] [PubMed]

95. Shi, L.; Gao, X.; Li, X.; Jiang, N.; Luo, F.; Gu, C.; Chen, M.; Cheng, H.; Liu, P. Ellagic acid enhances the efficacy of PI3K inhibitor GDC-0941 in breast cancer cells. *Curr. Mol. Med.* **2015**, *15*, 478–486. [CrossRef] [PubMed]

96. Barrajón-Catalán, E.; Fernández-Arroyo, S.; Saura, D.; Guillén, E.; Fernández-Gutiérrez, A.; Segura-Carretero, A.; Micol, V. Cistaceae aqueous extracts containing ellagitannins show antioxidant and antimicrobial capacity, and cytotoxic activity against human cancer cells. *Food Chem. Toxicol.* **2010**, *48*, 2273–2282. [CrossRef] [PubMed]

97. Miyamoto, K.I.; Nomura, M.; Sasakura, M.; Matsui, E.; Koshiura, R.; Murayama, T.; Furukawa, T.; Hatano, T.; Yoshida, T.; Okuda, T. Antitumor activity of oenothein B, a unique macrocyclic ellagitannin. *Jpn. J. Cancer Res. Gann* **1993**, *84*, 99–103. [CrossRef] [PubMed]

98. Enzinger, P.C.; Mayer, R.J. Esophageal cancer. *N. Engl. J. Med.* **2003**, *349*, 2241–2252. [CrossRef] [PubMed]

99. De Stefani, E.; Barrios, E.; Fierro, L. Black (air-cured) and blond (flue-cured) tobacco and cancer risk. III: Oesophageal cancer. *Eur. J. Cancer* **1993**, *29A*, 763–766. [CrossRef]

100. Stoner, G.D.; Chen, T.; Kresty, L.A.; Aziz, R.M.; Reinemann, T.; Nines, R. Protection against esophageal cancer in rodents with lyophilized berries: Potential mechanisms. *Nutr. Cancer* **2006**, *54*, 33–46. [CrossRef] [PubMed]

101. Kresty, L.A.; Morse, M.A.; Morgan, C.; Carlton, P.S.; Lu, J.; Gupta, A.; Blackwood, M.; Stoner, G.D. Chemoprevention of esophageal tumorigenesis by dietary administration of lyophilized black raspberries. *Cancer Res.* **2001**, *61*, 6112–6119. [PubMed]

102. Bishayee, A.; Haskell, Y.; Do, C.; Siveen, K.S.; Mohandas, N.; Sethi, G.; Stoner, G.D. Potential benefits of edible berries in the management of aerodigestive and gastrointestinal tract cancers: Preclinical and clinical evidence. *Crit. Rev. Food Sci. Nutr.* **2015**. in press. [CrossRef] [PubMed]

103. Mandal, S.; Stoner, G.D. Inhibition of *N*-nitrosobenzylmethylamine-induced esophageal tumorigenesis in rats by ellagic acid. *Carcinogenesis* **1990**, *11*, 55–61. [CrossRef] [PubMed]

104. Daniel, E.M.; Stoner, G.D. The effects of ellagic acid and 13-cis-retinoic acid on *N*-nitrosobenzylmethylamine-induced esophageal tumorigenesis in rats. *Cancer Lett.* **1991**, *56*, 117–124. [CrossRef]

105. Stoner, G.D.; Morse, M.A. Isothiocyanates and plant polyphenols as inhibitors of lung and esophageal cancer. *Cancer Lett.* **1997**, *114*, 113–119. [CrossRef]

106. Wang, L.S.; Dombkowski, A.A.; Seguin, C.; Rocha, C.; Cukovic, D.; Mukundan, A.; Henry, C.; Stoner, G.D. Mechanistic basis for the chemopreventive effects of black raspberries at a late stage of rat esophageal carcinogenesis. *Mol. Carcinog.* **2011**, *50*, 291–300. [CrossRef] [PubMed]

107. Wang, L.S.; Hecht, S.; Carmella, S.; Seguin, C.; Rocha, C.; Yu, N.; Stoner, K.; Chiu, S.; Stoner, G. Berry ellagitannins may not be sufficient for prevention of tumors in the rodent esophagus. *J. Agric. Food Chem.* **2010**, *58*, 3992–3995. [CrossRef] [PubMed]

108. Liu, H.; Li, J.; Zhao, W.; Bao, L.; Song, X.; Xia, Y.; Wang, X.; Zhang, C.; Wang, X.; Yao, X. Fatty acid synthase inhibitors from *Geum japonicum* Thunb. var. chinense. *Chem. Biodivers.* **2009**, *6*, 402–410. [CrossRef] [PubMed]

109. Zhang, Y.; Seeram, N.P.; Lee, R.; Feng, L.; Heber, D. Isolation and identification of strawberry phenolics with antioxidant and human cancer cell antiproliferative properties. *J. Agric. Food Chem.* **2008**, *56*, 670–675. [CrossRef] [PubMed]

110. Weisburg, J.H.; Schuck, A.G.; Reiss, S.E.; Wolf, B.J.; Fertel, S.R.; Zuckerbraun, H.L.; Babich, H. Ellagic acid, a dietary polyphenol, selectively cytotoxic to HSC-2 oral carcinoma cells. *Anticancer Res.* **2013**, *33*, 1829–1836. [PubMed]

111. Zhu, X.; Xiong, L.; Zhang, X.; Shi, N.; Zhang, Y.; Ke, J.; Sun, Z.; Chen, T. Lyophilized strawberries prevent 7, 12-dimethylbenz [α] anthracene (DMBA)-induced oral squamous cell carcinogenesis in hamsters. *J. Funct. Foods* **2015**, *15*, 476–486. [CrossRef]

112. Casto, B.C.; Knobloch, T.J.; Galioto, R.L.; Yu, Z.; Accurso, B.T.; Warner, B.M. Chemoprevention of oral cancer by lyophilized strawberries. *Anticancer Res.* **2013**, *33*, 4757–4766. [PubMed]
113. Priyadarsini, R.V.; Kumar, N.; Khan, I.; Thiyagarajan, P.; Kondaiah, P.; Nagini, S. Gene expression signature of DMBA-induced hamster buccal pouch carcinomas: Modulation by chlorophyllin and ellagic acid. *PLoS ONE* **2012**, *7*, e34628. [CrossRef] [PubMed]
114. Anitha, P.; Priyadarsini, R.V.; Kavitha, K.; Thiyagarajan, P.; Nagini, S. Ellagic acid coordinately attenuates Wnt/β-catenin and NF-κb signaling pathways to induce intrinsic apoptosis in an animal model of oral oncogenesis. *Eur. J. Nutr.* **2013**, *52*, 75–84. [CrossRef] [PubMed]
115. Kowshik, J.; Giri, H.; Kranthi Kiran Kishore, T.; Kesavan, R.; Naik Vankudavath, R.; Bhanuprakash Reddy, G.; Dixit, M.; Nagini, S. Ellagic acid inhibits VEGF/VEGFR2, PI3K/Akt and MAPK signaling cascades in the hamster cheek pouch carcinogenesis model. *Anti-Cancer Agents Med. Chem.* **2014**, *14*, 1249–1260. [CrossRef]
116. Ding, Y.; Yao, H.; Yao, Y.; Fai, L.Y.; Zhang, Z. Protection of dietary polyphenols against oral cancer. *Nutrients* **2013**, *5*, 2173–2191. [CrossRef] [PubMed]
117. Naghavi, M.; Wang, H.; Lozano, R.; Davis, A.; Liang, X.; Zhou, M.; Vollset, S.E.; Ozgoren, A.A.; Abdalla, S.; Abd-Allah, F. Global, regional, and national age-sex specific all-cause and cause-specific mortality for 240 causes of death, 1990–2013: A systematic analysis for the global burden of disease study 2013. *Lancet* **2015**, *385*, 117–171.
118. Oh, G.-S.; Pae, H.-O.; Oh, H.; Hong, S.-G.; Kim, I.-K.; Chai, K.-Y.; Yun, Y.-G.; Kwon, T.-O.; Chung, H.-T. *In vitro* anti-proliferative effect of 1,2,3,4,6-penta-*O*-galloyl-beta-D-glucose on human hepatocellular carcinoma cell line, SK-HEP-1 cells. *Cancer Lett.* **2001**, *174*, 17–24. [CrossRef]
119. Yin, S.; Dong, Y.; Li, J.; Lü, J.; Hu, H. Penta-1,2,3,4,6-*O*-galloyl-beta-D-glucose induces senescence-like terminal S-phase arrest in human hepatoma and breast cancer cells. *Mol. Carcinog.* **2011**, *50*, 592–600. [CrossRef] [PubMed]
120. Dong, Y.; Yin, S.; Jiang, C.; Luo, X.; Guo, X.; Zhao, C.; Fan, L.; Meng, Y.; Lu, J.; Song, X. Involvement of autophagy induction in penta-1,2,3,4,6-*O*-galloyl-β-D-glucose-induced senescence-like growth arrest in human cancer cells. *Autophagy* **2014**, *10*, 296–310. [CrossRef] [PubMed]
121. Hau, D.K.-P.; Zhu, G.-Y.; Leung, A.K.-M.; Wong, R.S.-M.; Cheng, G.Y.-M.; Lai, P.B.; Tong, S.-W.; Lau, F.-Y.; Chan, K.-W.; Wong, W.-Y.; et al. *In vivo* anti-tumour activity of corilagin on Hep3B hepatocellular carcinoma. *Phytomedicine* **2010**, *18*, 11–15. [CrossRef] [PubMed]
122. Ming, Y.; Zheng, Z.; Chen, L.; Zheng, G.; Liu, S.; Yu, Y.; Tong, Q. Corilagin inhibits hepatocellular carcinoma cell proliferation by inducing G2/M phase arrest. *Cell Biol. Int.* **2013**, *37*, 1046–1054. [CrossRef] [PubMed]
123. Zhang, T.-T.; Yang, L.; Jiang, J.-G. Effects of thonningianin A in natural foods on apoptosis and cell cycle arrest of HepG-2 human hepatocellular carcinoma cells. *Food Funct.* **2015**, *6*, 2588–2597. [CrossRef] [PubMed]
124. CDC. Cervical Cancer Statistics. Available online: http://www.cdc.gov/cancer/cervical/statistics/ (accessed on 21 March 2016).
125. Bosch, F.X.; Manos, M.M.; Munoz, N.; Sherman, M.; Jansen, A.M.; Peto, J.; Schiffman, M.H.; Moreno, V.; Kurman, R.; Shah, K.V.; et al. Prevalence of human papillomavirus in cervical cancer: A worldwide perspective. *J. Natl. Cancer Inst.* **1995**, *87*, 796–802. [CrossRef] [PubMed]
126. Ramasamy, S.; Abdul Wahab, N.; Zainal Abidin, N.; Manickam, S.; Zakaria, Z. Growth inhibition of human gynecologic and colon cancer cells by phyllanthus watsonii through apoptosis induction. *PLoS ONE* **2012**, *7*, e34793.
127. Ross, H.A.; McDougall, G.J.; Stewart, D. Antiproliferative activity is predominantly associated with ellagitannins in raspberry extracts. *Phytochemistry* **2007**, *68*, 218–228. [CrossRef] [PubMed]
128. Yi, Z.C.; Liu, Y.Z.; Li, H.X.; Yin, Y.; Zhuang, F.Y.; Fan, Y.B.; Wang, Z. Tellimagrandin I enhances gap junctional communication and attenuates the tumor phenotype of human cervical carcinoma HeLa cells *in vitro*. *Cancer Lett.* **2006**, *242*, 77–87. [CrossRef] [PubMed]
129. Wang, C.C.; Chen, L.G.; Yang, L.L. Camelliin B induced apoptosis in HeLa cell line. *Toxicology* **2001**, *168*, 231–240. [CrossRef]
130. Le, V.; Esposito, D.; Grace, M.H.; Ha, D.; Pham, A.; Bortolazzo, A.; Bevens, Z.; Kim, J.; Okuda, R.; Komarnytsky, S.; et al. Cytotoxic effects of ellagitannins isolated from walnuts in human cancer cells. *Nutr. Cancer* **2014**, *66*, 1304–1314. [CrossRef] [PubMed]

131. Moktar, A.; Ravoori, S.; Vadhanam, M.V.; Gairola, C.G.; Gupta, R.C. Cigarette smoke-induced DNA damage and repair detected by the comet assay in HPV-transformed cervical cells. *Int. J. Oncol.* **2009**, *35*, 1297–1304. [PubMed]
132. Khan, N.; Afaq, F.; Kweon, M.-H.; Kim, K.; Mukhtar, H. Oral consumption of pomegranate fruit extract inhibits growth and progression of primary lung tumors in mice. *Cancer Res.* **2007**, *67*, 3475–3482. [CrossRef] [PubMed]
133. Khan, N.; Hadi, N.; Afaq, F.; Syed, D.N.; Kweon, M.H.; Mukhtar, H. Pomegranate fruit extract inhibits prosurvival pathways in human A549 lung carcinoma cells and tumor growth in athymic nude mice. *Carcinogenesis* **2007**, *28*, 163–173. [CrossRef] [PubMed]
134. Zahin, M.; Ahmad, I.; Gupta, R.C.; Aqil, F. Punicalagin and ellagic acid demonstrate antimutagenic activity and inhibition of benzo [a] pyrene induced DNA adducts. *BioMed Res. Int.* **2014**, *2014*. [CrossRef] [PubMed]
135. Kulkarni, A.P.; Mahal, H.; Kapoor, S.; Aradhya, S. *In vitro* studies on the binding, antioxidant, and cytotoxic actions of punicalagin. *J. Agric. Food chem.* **2007**, *55*, 1491–1500. [CrossRef] [PubMed]
136. Kuo, P.-L.; Hsu, Y.-L.; Lin, T.-C.; Lin, L.-T.; Chang, J.-K.; Lin, C.-C. Casuarinin from the bark of *Terminalia arjuna* induces apoptosis and cell cycle arrest in human breast adenocarcinoma MCF-7 cells. *Planta Med.* **2005**, *71*, 237–243. [CrossRef] [PubMed]
137. Yoshimura, M.; Watanabe, Y.; Kasai, K.; Yamakoshi, J.; Koga, T. Inhibitory effect of an ellagic acid-rich pomegranate extract on tyrosinase activity and ultraviolet-induced pigmentation. *Biosci. Biotechnol. Biochem.* **2005**, *69*, 2368–2373. [CrossRef] [PubMed]
138. Afaq, F.; Zaid, M.A.; Khan, N.; Dreher, M.; Mukhtar, H. Protective effect of pomegranate-derived products on UVB-mediated damage in human reconstituted skin. *Exp. Dermatol.* **2009**, *18*, 553–561. [CrossRef] [PubMed]
139. Afaq, F.; Zaid, M.; Khan, N.; Syed, D.; Yun, J.-M.; Sarfaraz, S.; Suh, Y.; Mukhtar, H. Inhibitory effect of oral feeding of pomegranate fruit extract on UVB-induced skin carcinogenesis in SKH-1 hairless mice. In Proceedings of the 99th AACR Annual Meeting, San Diego, CA, USA, 12–16 April 2008; AACR Publications: Philadelphia, PA, USA; San Diego, CA, USA; p. 1246.
140. Chung, K.-T.; Wei, C.-I.; Johnson, M.G. Are tannins a double-edged sword in biology and health? *Trends Food Sci. Technol.* **1998**, *9*, 168–175. [CrossRef]
141. Mennen, L.I.; Walker, R.; Bennetau-Pelissero, C.; Scalbert, A. Risks and safety of polyphenol consumption. *Am. J. Clin. Nutr.* **2005**, *81*, 326S–329S. [PubMed]
142. Sánchez-Lamar, A.; Fonseca, G.; Fuentes, J.L.; Cozzi, R.; Cundari, E.; Fiore, M.; Ricordy, R.; Perticone, P.; Degrassi, F.; de Salvia, R. Assessment of the genotoxic risk of *Punica granatum* L.(Punicaceae) whole fruit extracts. *J. Ethnopharmacol.* **2008**, *115*, 416–422. [CrossRef] [PubMed]
143. Labieniec, M.; Gabryelak, T. Effects of tannins on Chinese hamster cell line B14. *Mutat. Res. Genet. Toxicol. Environ. Mutagen.* **2003**, *539*, 127–135. [CrossRef]
144. Chen, S.C.; Chung, K.T. Mutagenicity and antimutagenicity studies of tannic acid and its related compounds. *Food Chem. Toxicol.* **2000**, *38*, 1–5. [CrossRef]
145. Filippich, L.J.; Zhu, J.; Oelrichs, P.; Alsalami, M.T.; Doig, A.J.; Cao, G.R.; English, P.B. Hepatotoxic and nephrotoxic principles in terminalia oblongata. *Res. Vet. Sci.* **1991**, *50*, 170–177. [CrossRef]
146. Cerdá, B.; Cerón, J.J.; Tomás-Barberán, F.A.; Espín, J.C. Repeated oral administration of high doses of the pomegranate ellagitannin punicalagin to rats for 37 days is not toxic. *J. Agric. Food Chem.* **2003**, *51*, 3493–3501. [CrossRef] [PubMed]
147. McDougall, G.J.; Shpiro, F.; Dobson, P.; Smith, P.; Blake, A.; Stewart, D. Different polyphenolic components of soft fruits inhibit α-amylase and α-glucosidase. *J. Agric. Food Chem.* **2005**, *53*, 2760–2766. [CrossRef] [PubMed]
148. Godbout, A.; Chiasson, J.L. Who should benefit from the use of alpha-glucosidase inhibitors? *Curr. Diabetes Rep.* **2007**, *7*, 333–339. [CrossRef]
149. Li, H.; Tanaka, T.; Zhang, Y.-J.; Yang, C.-R.; Kouno, I. Rubusuaviins A–F, monomeric and oligomeric ellagitannins from Chinese sweet tea and their α-amylase inhibitory activity. *Chem. Pharm. Bull.* **2007**, *55*, 1325–1331. [CrossRef] [PubMed]
150. Santos-Buelga, C.; Scalbert, A. Proanthocyanidins and tannin-like compounds—Nature, occurrence, dietary intake and effects on nutrition and health. *J. Sci. Food Agric.* **2000**, *80*, 1094–1117. [CrossRef]

151. Frutos, P.; Raso, M.; Hervás, G.; Mantecón, Á.R.; Pérez, V.; Giráldez, F.J. Is there any detrimental effect when a chestnut hydrolysable tannin extract is included in the diet of finishing lambs? *Anim. Res.* **2004**, *53*, 127–136. [CrossRef]

152. Tasaki, M.; Umemura, T.; Maeda, M.; Ishii, Y.; Okamura, T.; Inoue, T.; Kuroiwa, Y.; Hirose, M.; Nishikawa, A. Safety assessment of ellagic acid, a food additive, in a subchronic toxicity study using F344 rats. *Food Chem. Toxicol.* **2008**, *46*, 1119–1124. [CrossRef] [PubMed]

153. Patel, C.; Dadhaniya, P.; Hingorani, L.; Soni, M. Safety assessment of pomegranate fruit extract: Acute and subchronic toxicity studies. *Food Chem. Toxicol.* **2008**, *46*, 2728–2735. [CrossRef] [PubMed]

154. Murphy, M.M.; Barraj, L.M.; Spungen, J.H.; Herman, D.R.; Randolph, R.K. Global assessment of select phytonutrient intakes by level of fruit and vegetable consumption. *Br. J. Nutr.* **2014**, *112*, 1004–1018. [CrossRef] [PubMed]

155. Ovaskainen, M.-L.; Törrönen, R.; Koponen, J.M.; Sinkko, H.; Hellström, J.; Reinivuo, H.; Mattila, P. Dietary intake and major food sources of polyphenols in Finnish adults. *J. Nutr.* **2008**, *138*, 562–566. [PubMed]

156. Radtke, J.; Linseisen, J.; Wolfram, G. Phenolic acid intake of adults in a Bavarian subgroup of the national food consumption survey. *Z. Ernahrungswiss.* **1998**, *37*, 190–197. [CrossRef] [PubMed]

toxins

MDPI

Article

Cancer Therapy by Catechins Involves Redox Cycling of Copper Ions and Generation of Reactive Oxygenspecies

Mohd Farhan [1], Husain Yar Khan [2], Mohammad Oves [3], Ahmed Al-Harrasi [2], Nida Rehmani [1], Hussain Arif [1], Sheikh Mumtaz Hadi [1,*] and Aamir Ahmad [4,*]

[1] Department of Biochemistry, Faculty of Life Sciences, AMU, Aligarh 202002, India;
    farhan@mohdfarhan.com (M.F.); nida.rehmani4@gmail.com (N.R.); arifkap@gmail.com (H.A.)
[2] UoN Chair of Oman's Medicinal Plants and Marine Natural Products, University of Nizwa, Birkat Al Mauz,
    PO Box 33, Postal Code 616, Nizwa, Oman; husainyar@gmail.com (H.Y.K.);
    aharrasi@unizwa.edu.om (A.A.-H.)
[3] Center of Excellence in Environmental Studies, King Abdulaziz University, Jeddah 21589, Saudia Arabia;
    owais.micro@gmail.com
[4] Karmanos Cancer Institute and Wayne State School of Medicine, Detroit, MI 48201, USA
*   Correspondence: ahmada@karmanos.org (A.A.); smhadi1946@gmail.com (S.M.H.);
    Tel.: +1-313-576-8315 (A.A.); +91-983-726-6761 (S.M.H.); Fax: +1-313-576-8389 (A.A.)

Academic Editor: Carmela Fimognari
Received: 2 January 2016; Accepted: 26 January 2016; Published: 4 February 2016

Abstract: Catechins, the dietary phytochemicals present in green tea and other beverages, are considered to be potent inducers of apoptosis and cytotoxicity to cancer cells. While it is believed that the antioxidant properties of catechins and related dietary agents may contribute to lowering the risk of cancer induction by impeding oxidative injury to DNA, these properties cannot account for apoptosis induction and chemotherapeutic observations. Catechin (C), epicatechin (EC), epigallocatechin (EGC) and epigallocatechin-3-gallate (EGCG) are the four major constituents of green tea. In this article, using human peripheral lymphocytes and comet assay, we show that C, EC, EGC and EGCG cause cellular DNA breakage and can alternatively switch to a prooxidant action in the presence of transition metals such as copper. The cellular DNA breakage was found to be significantly enhanced in the presence of copper ions. Catechins were found to be effective in providing protection against oxidative stress induced by tertbutylhydroperoxide, as measured by oxidative DNA breakage in lymphocytes. The prooxidant action of catechins involved production of hydroxyl radicals through redox recycling of copper ions. We also determined that catechins, particularly EGCG, inhibit proliferation of breast cancer cell line MDA-MB-231 leading to a prooxidant cell death. Since it is well established that tissue, cellular and serum copper levels are considerably elevated in various malignancies, cancer cells would be more subject to redox cycling between copper ions and catechins to generate reactive oxygen species (ROS) responsible for DNA breakage. Such a copper dependent prooxidant cytotoxic mechanism better explains the anticancer activity and preferential cytotoxicity of dietary phytochemicals against cancer cells.

Keywords: catechins; prooxidant; anticancer; copper; DNA breakage; reactive oxygen species; epicatechin; epigallocatechin; epigallocatechin-3-gallate

1. Introduction

In recent years, there has been an increasing interest in understanding the potential of cancer chemopreventive properties of plant derived polyphenolic compounds. Epidemiological evidence suggests that a high consumption of dietary products derived from plant sources among certain populations may reduce their risk of cancer induction as compared to those with low intakes [1–3]. This has been associated with human food stuff found to be rich in a wide variety of biologically active compounds [4]. Among such dietary constituents, catechins (a major component in green tea) are considered to be the most effective in cancer chemoprevention in humans. Catechins are a sub class of plant polyphenols that particularly include (+)-catechin (C), (−)-epicatechin (EC), (−)-epigallocatechin (EGC), and (−)-epigallocatechin-3-gallate (EGCG) and are considered as the most effective cytotoxic agents and inducers of apoptosis in cancer cells [5–8]. In recent years, several reports have documented that plant polyphenols including catechins induce apoptosis in various cell lines [9–13]. Of particular interest is the observation that a number of these polyphenols including catechins induce apoptotic cell death in various cell lines but not in normal cells [5,10–12,14]. However, the mechanism by which these compounds inhibit cell proliferation and induce apoptosis in cancer cells has been the subject of much interest. These compounds possess both antioxidant as well as prooxidant properties [6,7,15]. Evidence in the literature suggests that the antioxidant properties of such plant polyphenols may not fully account for their observed antiproliferative and cancer therapeutic effects [16]. We earlier proposed a hypothesis where we had explained that the prooxidant, rather than the antioxidant, activity of these compounds is important for their anticancer effects [17–19]. Such a prooxidant effect is induced in the presence of transition metals, such as copper. Copper is an important metal ion present in chromatin and is closely associated with DNA bases, particularly guanine [20]. It is one of the most redox active among the various metal ions present in biological systems facilitating rapid recycling, in the presence of molecular oxygen and compounds such as plant polyphenols, leading to the formation of reactive oxygen species (ROS) such as the hydroxyl radical.

In this article, we show that catechins can alternately behave as prooxidants in the presence of Cu(II) leading to cytotoxic action. Identification of molecular targets, modulation of which is associated with inhibition of malignantly transformed cells, is vital to cancer prevention and will greatly assist in a better understanding of anticancer mechanisms by naturally occurring chemotherapeutic compounds. A number of reports in the literature have established that tissue, serum, and cellular copper levels are considerably elevated in various malignancies [21–24]. Therefore, cancer cells may be more subject to electron transfer between copper ions and these catechins to generate ROS [25,26]. Such a mechanism, which involves a copper-dependent pathway of cell death, better explains the anticancer properties of polyphenols of diverse chemical structures, as also their preferential toxicity against cancer cells. The structures of various catechins used in this study are shown in Figure 1.

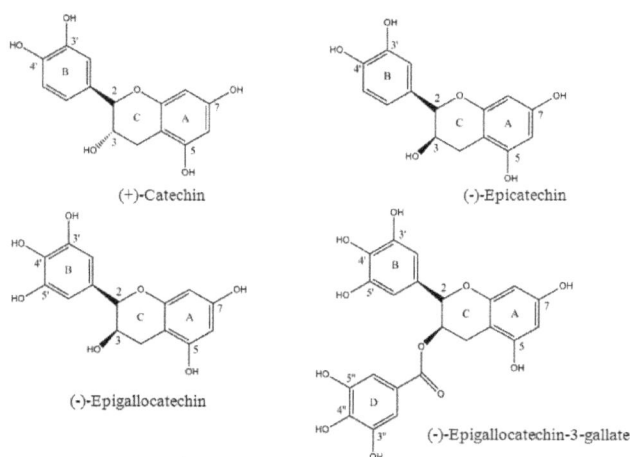

Figure 1. Chemical structure of catechin, epicatechin, epigallocatechin and epigallocatechin-3-gallate.

2. Results

2.1. Formation of Catechins-Cu(II) Complex

The possibility for the formation of C/EC/EGC/EGCG with Cu(II) complex was examined. This was carried out by recording the absorption spectra of C, EC, EGC and EGCG with increasing concentrations of Cu(II). The results given in Figure 2 show that the addition of Cu(II) to C, EC, EGC and EGCG results in an enhancement in the peak appearing at their respective λ_{max}. The absorption spectra of C, EC, EGC and EGCG in the presence of copper suggests a simple mode of interaction between these catechins and Cu(II). The absorption maxima of C, EC, EGC and EGCG lie in the range of 260–280 nm.

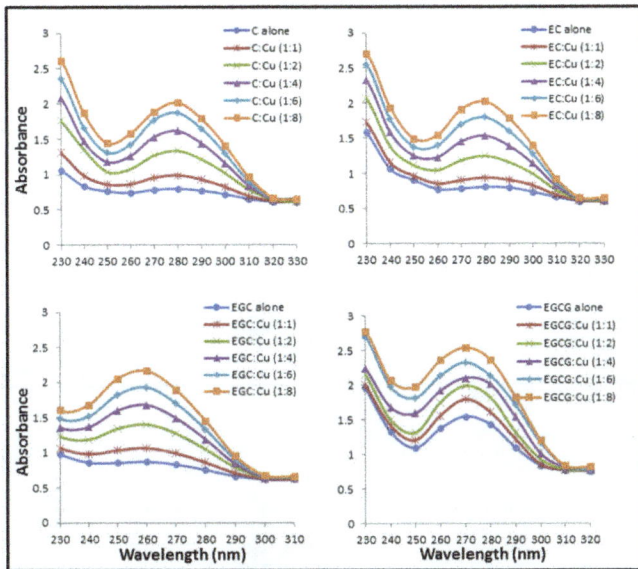

Figure 2. Effect of increasing copper concentrations on the absorbance spectra of Catechin (C), Epicatechin (EC), Epigallocatechin (EGC) and Epigallocatechin-3-gallate (EGCG). Catechins (in 10 mM Tris-HCl, pH 7.5) absorption spectra were recorded in the presence of increasing concentration of Cu(II).

2.2. Formation of Complexes Involving Cu(II) with Catechins

Figure 3 shows the effect of addition of increasing molar base pair ratios of Cu(II) on the fluorescence emission spectra of C, EC, EGC and EGCG excited at 273 nm (approximate absorption maximum of catechins). The result shown in Figure 3 indicates binding as addition of Cu(II) causes quenching of C, EC, EGC and EGCG fluorescence. These results support the result of absorption studies shown in Figure 2 where formation of catechins-copper complex was demonstrated.

2.3. Detection of Catechins Induced Cu(I) Production by Bathocuproine

The production of Cu(I), formed as a result of reduction of Cu(II) by C, EC, EGC and EGCG, was analyzed using bathocuproine which is a selective Cu(I) sequestering agent that binds specifically to the reduced form of copper, *i.e.*, Cu(I), but not to the oxidized form [27]. The Cu(I)-chelates exhibit an absorption maximum at 480 nm. As shown in Figure 4, Cu(II) does not interfere with the maxima, whereas C + Cu(II), EC + Cu(II), EGC + Cu(II) and EGCG + Cu(II) react to generate Cu(I) which complexes with bathocuproine to give a peak appearing at 480 nm. The results show that these catechins are able to reduce Cu(II) to Cu(I) and contribute to the redox cycling of the metal.

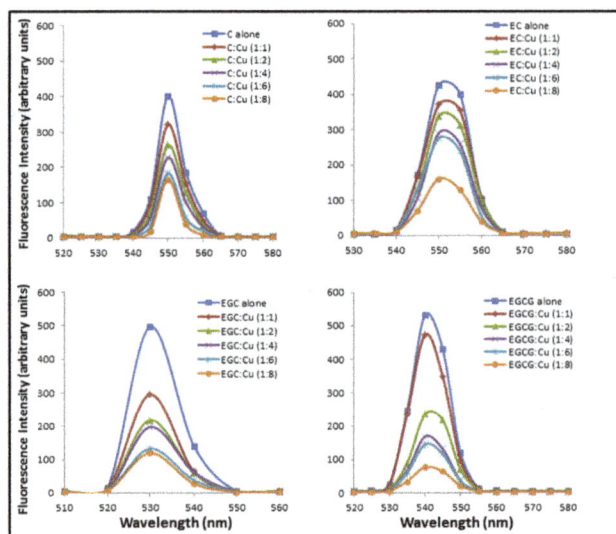

Figure 3. Effect of increasing copper concentrations on the fluorescence emission spectra of C, EC, EGC and EGCG. Catechins (in 10 mM Tris-HCl, pH 7.5) were excited at 273 nm in the presence of increasing concentration of Cu(II) and the emission spectra were recorded between 510 and 580 nm.

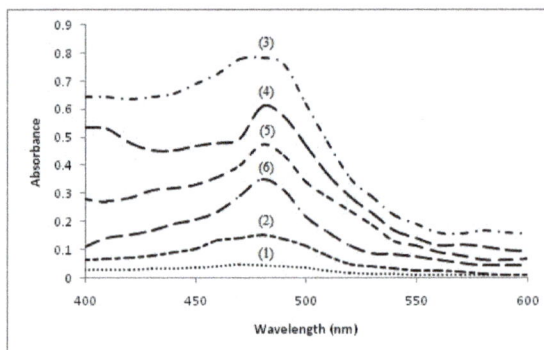

Figure 4. Detection of catechin induced Cu(I) production by Bathocuproine. Reaction mixture (3.0 mL) contained 3.0 mMTris-HCl (pH 7.5) along with 300 μM bathocuproine and indicated concentrations of the following: (1) Bathocuproine + 100 μM Cu(II); (2) Bathocuproine + 100 μM Cu(I); (3) Bathocuproine + 50 μM C + 100 μM Cu(II); (4) Bathocuproine + 50 μM EC + 100 μM Cu(II); (5) Bathocuproine + 50 μM EGC + 100 μM Cu(II); and (6) Bathocuproine + 50 μM EGCG + 100 μM Cu(II). The Bathocuproine alone or Bathocuproine in the presence of respective compounds did not interfere with the Bathocuproine-Cu(I) complex peak at 480 nm (not shown).

2.4. Superoxide Production by Catechins

The production of superoxide anion was determined by the Nakayama method [28], which involves reduction of NBT by C, EC, EGC and EGCG to a formazan. The time dependent generation of superoxide anion by C, EC, EGC and EGCG as evidenced by the increase in absorbance at 560 nm is shown in Figure 5. The fact that NBT was genuinely assaying superoxide was confirmed by SOD (100 μg/mL) inhibiting the reaction (results not shown). It is known that superoxide may undergo automatic dismutation to form H_2O_2 which in the presence of transition metals such as copper favors Fenton type reaction to generate hydroxyl radicals which could act as a proximal DNA cleaving agent leading to oxidative DNA breakage.

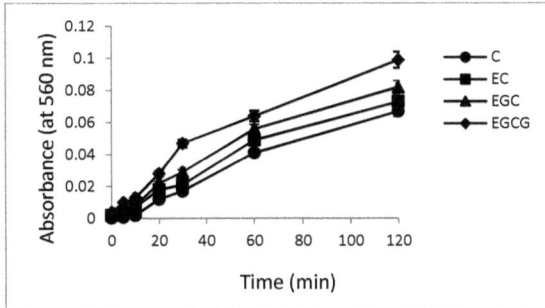

Figure 5. Photogeneration of superoxide anion by catechins on illumination under fluorescent light. Reaction mixture contained 50 mM phosphate buffer (pH 7.5) and 50 μM of C, EC, EGC and EGCG. The samples were placed at a distance of 10 cm from the light source. All values reported are means of three independent experiments. Error bars represent standard error of mean.

2.5. Hydroxyl Radical Generation by Catechins

It has been previously shown that during the reduction of Cu(II) to Cu(I), reactive oxygen species such as hydroxyl radicals are formed which serve as the proximal DNA cleaving agent [29]. Therefore, the capacity of C, EC, EGC and EGCG to generate hydroxyl radical in the presence of Cu(II) was examined. The assay is based on the fact that degradation of DNA by hydroxyl radicals results in the release of TBA (2-thiobarbituric acid) reactive material, which forms a colored adduct with TBA whose absorbance is read at 532 nm [30]. The results given in Figure 6 clearly show that increasing concentrations of catechins lead to a progressive increase in the formation of hydroxyl radicals.

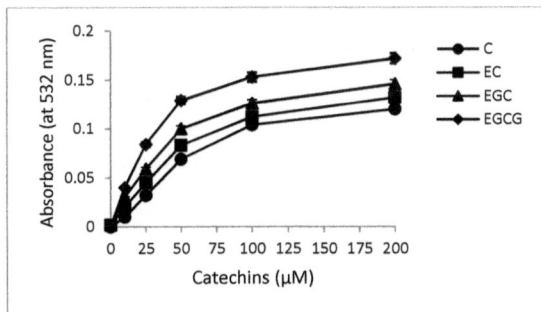

Figure 6. Formation of hydroxyl radicals as a function of catechins concentration in the presence of Cu(II). Reaction mixture (0.5 mL) contained 200 μg calf thymus DNA as substrate, 100 μM Cu(II) and indicated concentrations of C, EC, EGC and EGCG. The reaction mixture was incubated at 37 °C for 1 h. Hydroxyl radical formation was measured by determining the TBA reactive material. All values reported are means of three independent experiments. Error bars represent standard error of mean.

2.6. Breakage of Calf Thymus DNA by Catechins in the Presence of Cu(II)

C, EC, EGC and EGCG in the presence of Cu(II) were found to generate single strand specific nuclease sensitive sites in calf thymus DNA. The reaction was assessed by recording the proportion of DNA converted to acid soluble-nucleotides by the nuclease. Table 1 gives the dose response data of such a reaction. However, C, EC, EGC and EGCG in the absence of Cu(II) did not show appreciable degradation of calf thymus DNA. Control experiments (data not shown) established that heat denatured DNA underwent 100% hydrolysis following the treatment with nuclease. In the presence of Cu(II) (50 μM), increasing concentrations of C, EC, EGC and EGCG resulted in an increase in nuclease sensitive sites in DNA leading to increased DNA hydrolysis.

Table 1. Degradation of calf thymus DNA by the catechins in the presence of Cu(II) as measured by the degree of single strand specific S_1-nucleasedigestion. Reaction mixture (0.5 mL) containing 10 mM Tris-HCl (pH 7.5) and 500 μg calf thymus DNA was incubated at 37 °C with indicated concentrations of respective polyphenol alone or polyphenol with Cu(II) (100 μM). All values represent mean ± SEM of three independent experiments.

Catechins	Concentration (μM)	% DNA Hydrolyzed	
		without Cu(II)	with Cu(II)
C	50	1.53 ± 0.21	8.87 ± 0.46
	100	3.09 ± 0.37	11.93 ± 0.62
	200	5.98 ± 0.26	17.65 ± 0.73
	300	7.11 ± 0.67	26.34 ± 0.91
EC	50	2.62 ± 0.29	10.43 ± 0.41
	100	4.18 ± 0.35	15.07 ± 0.48
	200	7.26 ± 0.32	22.32 ± 0.81
	300	9.12 ± 0.56	28.56 ± 0.98
EGC	50	3.03 ± 0.43	12.71 ± 0.51
	100	5.21 ± 0.72	16.27 ± 0.81
	200	8.11 ± 0.54	22.45 ± 0.94
	300	10.87 ± 0.76	31.17 ± 1.13
EGCG	50	4.26 ± 0.43	14.11 ± 0.52
	100	8.67 ± 0.59	18.19 ± 0.77
	200	11.89 ± 0.74	29.58 ± 0.83
	300	14.35 ± 0.93	38.74 ± 1.71

2.7. Cleavage of Plasmid pBR322 DNA by Catechins

In order to examine the efficacy of catechins-Cu(II) system in DNA cleavage, as shown in Figure 7, we have tested the ability of C, EC, EGC and EGCG to cause cleavage of supercoiled plasmid pBR322 DNA in the presence of copper ions. As can be seen from the ethidium bromide stained agarose gel pattern, C, EC, EGC and EGCG alone show only some degree of DNA cleavage. However, addition of copper to these four catechins resulted in greater DNA cleavage, demonstrating that catechins are capable of plasmid DNA cleavage in the presence of copper ions.

2.8. Cellular DNA Breakage by catechins-Cu(II) in Lymphocytes as Measured by Comet Assay

We have earlier shown that most of the dietary polyphenolic phytochemicals, which are generally effective antioxidants, can switch to prooxidant action in the presence of transition metals such as copper [17,18]. In the experiment shown in Figure 8, we have tested the ability of C, EC, EGC and EGCG to cause DNA strand breaks in a cellular system of human peripheral lymphocytes both in the absence and the presence of Cu(II), as measured by standard comet assay. As seen from the figure, although all the compounds tested caused some breakage of cellular DNA, the degree of such breakage is enhanced in the presence of copper. Cu(II) (50 μM) controls were similar to untreated lymphocytes without any significant DNA breakage. The results clearly indicate that catechins-Cu(II) system is capable of DNA breakage in isolated lymphocytes and that such cellular DNA breakage is found of the order of EGCG > EGC > EC > C.

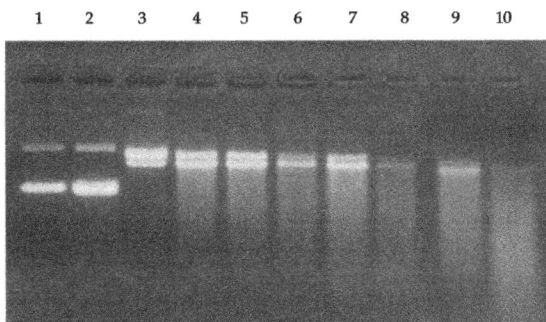

Figure 7. Agarose gel electrophoretic pattern of ethidium bromide stained pBR322 plasmid DNA after treatment with C, EC, EGC and EGCG in the absence and presence of copper. Lane 1: DNA alone; Lane 2: DNA + Cu(II) (50 μM); Lane 3: DNA + C (50 μM); Lane 4: DNA + C (50 μM) + Cu (II) (50 μM); Lane 5: DNA + EC (50 μM); Lane 6: DNA + EC (50 μM) + Cu (II) (50 μM); Lane 7: DNA + EGC (50 μM); Lane 8: DNA + EGC (50 μM) + Cu (II) (50 μM); Lane 9: DNA + EGCG (50 μM); Lane 10: DNA + EGCG (50 μM) + Cu (II) (50 μM).

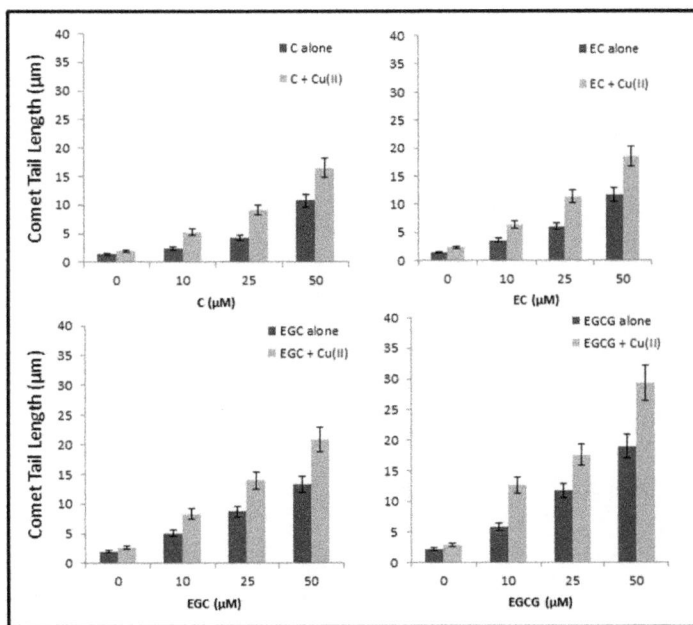

Figure 8. DNA breakage by catechins in human peripheral lymphocytes in the absence and presence of Cu(II). Comet tail length (μms) plotted as a function of increasing concentrations of catechins (0–50 μM) in the absence and presence of 50 μM Cu(II). All points represent mean of three independent experiments. Error bars denote Mean ± SEM. *p* value < 0.05 and significant when compared to control.

2.9. Determination of TBARS as a Measure of Oxidative Stress in Nuclei by Catechins in the Presence of Neocuproine and Thiourea

According to our hypothesis, the DNA breakage observed in lymphocyte nuclei is the result of the generation of hydroxyl radicals and other reactive oxygen species *in situ*. Oxygen radical damage to deoxyribose or DNA is considered to give rise to TBA reactive material [30,31]. We have therefore determined the formation of TBA reactive substance (TBARS) as a measure of oxidative stress in lymphocyte nuclei with increasing concentrations of C, EC, EGC and EGCG. The effect of pre-incubating the nuclei with neocuproine and thiourea was also studied. Results given in Figure 9 show a dose-dependent increase in the formation of TBA reactive substance in lymphocyte nuclei by C, EC, EGC and EGCG. However, a considerable decrease in the rate of formation of TBARS was observed in the presence of neocuproine and thiourea among all the four catechins used. The results indicate that DNA breakage in nuclei is inhibited by Cu (I) chelation and scavenging of reactive oxygen. Thus, it may be concluded that the oxidative stress induced by polyphenols in lymphocyte nuclei is at least in part mediated by chromatin bound copper.

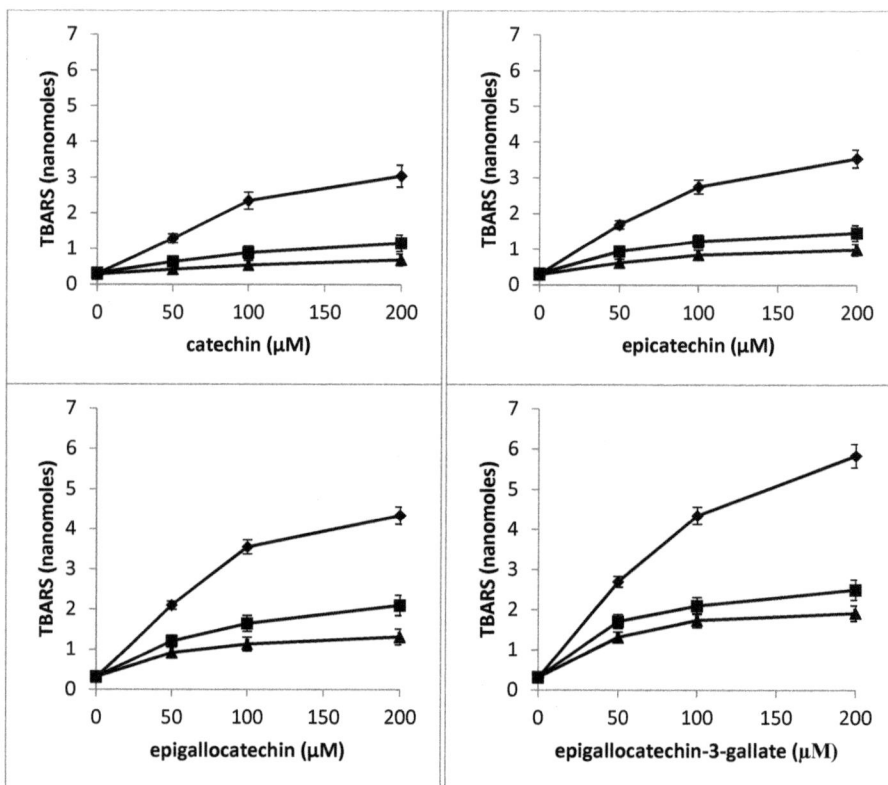

Figure 9. Effect of pre-incubation of lymphocyte nuclei with neocuproine and thiourea on TBARS generated by increasing concentrations of catechins: Catechin alone (filled circle), Catechin + neocuproine (1 mM) (filled square), and Catechin+thiourea (1 mM) (filled triangle). The nuclei suspension was pre-incubated with fixed concentration of neocuproine and thiourea for 30 min at 37 °C, after which it was further incubated for 1 h in the presence of increasing catechins concentration. Values reported are Mean ± SEM of three independent experiments.

2.10. Antioxidant Activity of Catechins against TBHP-Induced Oxidative Stress in Lymphocytes

TBHP is a well-known inducer of ROS-mediated oxidative stress that results in DNA damage [32,33]. In the present study, we have evaluated the antioxidant potential of C, EC, EGC and EGCG in providing protection to lymphocytes against TBHP induced oxidative injury. Figure 10 shows that whereas all the four catechins were able to inhibit the TBHP-induced lymphocyte DNA degradation, their relative antioxidant activities were different and appeared in the following order: EGCG > EGC > EC > C. The results indicate that EGCG is the most effective antioxidant among the four catechins used.

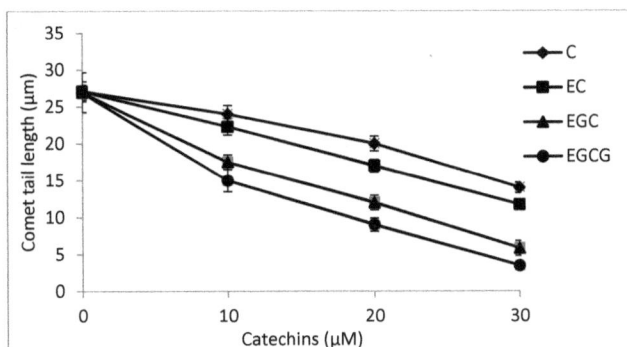

Figure 10. A comparison of antioxidant activities of various catechins as a function of decreasing tail length of comets against TBHP-induced oxidative DNA breakage in human peripheral lymphocytes as assessed by Comet assay. $p < 0.05$ by comparison with TBHP-treated positive control. Values reported are Mean $\pm$ SEM of three independent experiments.

2.11. Catechins Cause Inhibition of Cell Growth in MDA-MB-231 Breast Cancer Cells

In Figure 8, it was observed that catechins were able to cause strand breaks in cellular DNA. Subsequently, the effects of the various catechins were tested on the proliferative potential of human breast cancer MDA-MB-231 cells. As can be seen in Figure 11A, a dose-dependent inhibition of proliferation of breast cancer cells MDA-MB-231 by catechins was observed, as assessed by MTT assay. The order of activity was found to be EGCG > EGC > EC > C. These results complement the cellular DNA breakage studies. Further, we observed (Figure 11B) that the normal breast epithelial cells, MCF-10A, were quite resistant to EGCG treatment but their culture in copper-enriched medium resulted in sensitization to EGCG action ($p < 0.01$). These results are in agreement with our earlier published results [5] involving plant polyphenols.

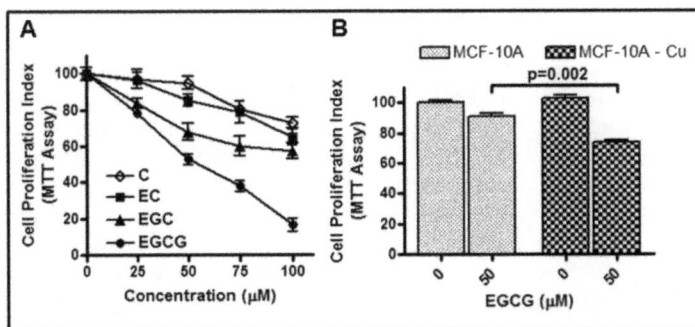

Figure 11. (**A**) The effects of C, EC, EGC and EGCG on the growth of MDA-MB-231 breast cancer cells as detected by MTT assay. The cells were incubated with indicated concentrations of catechins for 48 h, and the results are expressed relative to control (vehicle-treated) cells. (**B**) MCF10A (normal breast epithelial cells) and MCF10A+Cu (MCF-10A cells cultured in copper-enriched medium) were treated with either vehicle (0 μM) or 50 μM EGCG for 72 h.

3. Discussion

Studies mainly on anticancer mechanisms of plant polyphenols involve the induction of cell cycle arrest and modulation of transcription factors that lead to anti-neoplastic effects [10,34]. In light of the above findings in our laboratory and those of many others in the literature, it may be concluded that the plant polyphenols, particularly present in dietary agents, possessing anticancer and apoptosis-inducing activities are able to mobilize endogenous copper ions, possibly the copper bound to chromatin. Essentially, this would be an alternative, non-enzymatic, and copper-dependent pathway for the cytotoxic action of anticancer agents that are capable of mobilizing and reducing endogenous copper. As such, this would be independent of Fas and mitochondria mediated programmed cell deaths. It is conceivable that such a mechanism may also lead to internucleosomal DNA breakage (a hallmark of apoptosis), as internucleosomal spacer DNA would be relatively more susceptible to cleavage by ROS. Indeed, such a common mechanism better explains the anticancer effects of dietary molecules (*i.e.*, catechins) studied above with diverse chemical structures as also their preferential cytotoxicity toward cancer cells. The generation of hydroxyl radicals in the proximity of DNA is well established as a cause of strand scission. It is generally recognized that such a reaction with DNA is preceded by the association of a ligand with DNA followed by the formation of hydroxyl radicals at that site. The location of the redox-active metal is of utmost importance because the hydroxyl radical, owing to its extreme reactivity, interacts exclusively in the vicinity of the bound metal [35]. Copper ions are known to interact with both DNA phosphates and the bases through coordination binding [36]. Further, copper is also present in chromatin and is closely associated with DNA bases, particularly guanine [20]. Direct interaction of catechins with the DNA bound copper ions in a ternary complex and localized generation of non-diffusible hydroxyl radicals is a likely mechanism involved in catechin/Cu(II)-induced DNA cleavage. It has already been reported by us that the number of galloyl moieties present in catechinsplay an important role in cellular DNA breakage and catechins on reducing copper lead to the formation of "oxidized species" of the compounds. [26]. It may be presumed that the concentration of metals such as copper in cells is a decisive factor in driving antioxidant property of catechins toward their prooxidant action. As already mentioned, cancer cells are known to contain elevated levels of copper [37] and therefore may be more subject to electron transfer with catechins to generate ROS [25,26]. In normal cells, there exists a balance between the free radical generation and the antioxidant defense [38]. However, it has been clearly documented that tumor cells are under persistent oxidative stress and have an altered antioxidant system [39] and thus further ROS stress in malignant cells reaching a threshold level could result in apoptosis [37]. These observations further suggest that neoplastic cells may be more vulnerable to oxidative stress because they function with a heightened basal level of ROS because of increased rate of growth and metabolism [40]. Thus, in cancer cells, an enhanced exposure to ROS generated through the antioxidant/catechin-induced redox activity of endogenous copper can overwhelm the cells antioxidant capacity, leading to irreversible damage and apoptosis.

Beyond the preclinical findings, catechins such as EGCG show little promise as potent chemopreventive agents in the clinical settings, mainly owing to their inefficient systemic delivery and poor bioavailability. Catechins, like other polyphenols, are rapidly metabolized *in vivo*, resulting in a short systemic half-life and low plasma concentrations in the free form. For instance, it has been observed that the peak plasma concentrations reached up to 1.3 μM and 3.1 μM in healthy volunteers receiving 687.5 mg EGCG and 663.5 mg of ECG, respectively [41]. In this context, it needs to be emphasized that the significance of our work lies not so much in the potential therapeutic action of catechins against cancer, but in establishing a principle, namely that it is possible to mobilize elevated levels of endogenous copper in cancer cells by catechins to promote a pro-oxidant cell death. Once such a principle is established, catechins can serve as lead compounds to synthesize/formulate novel, anticancer drugs with superior bioavailability and extended systemic half-life. Presumably, such anticancer drugs would have a better therapeutic impact than the catechins *per se*.

Nevertheless, a previous study from our lab has shown that oral administration of copper to rats can induce a copper overload in their lymphocytes, rendering such isolated lymphocytes more susceptible to EGCG-induced pro-oxidant cellular DNA breakage [42]. This suggests that in cancer cells, where there are considerably higher levels of copper, the concentration of catechins required to elicit such a pro-oxidant cell death mechanism would be significantly lower. Therefore, based on these observations, it is clear that the pro-oxidant action of polyphenols is physiologically feasible provided an appropriate microenvironment with elevated copper levels is present, as is the scenario in cancer cells.

Further, it should be noted that in real life situation, catechins like EGCG are only one of the several polyphenols consumed as part of the diet. Since various other polyphenols present in diet, such as flavonoids and tannins are also active as pro-oxidants [17], their cumulative plasma concentration, bioavailability and anticancer effect should be much greater than a single polyphenol alone. In fact, a combination of EGCG with luteolin, has been found to be more effective than either of the polyphenol alone in inducing apoptosis in cancer cell lines *in vitro* and inhibition of tumor growth in nude mouse xenograft model [43].

4. Materials and Methods

4.1. Chemicals, Reagents and Cell Line

(+)-Catechin, (−)-epicatechin, (−)-epigallocatechin, (−)-epigallocatechin-3-gallate, calf thymus DNA, cupric chloride, neocuproine, thiourea, agarose, low melting point agarose, RPMI 1640, Triton X-100, Trypan blue, Histopaque1077, and phosphate buffered saline (PBS) Ca^{2+} and Mg^{2+} free were purchased from Sigma (St. Louis, MO, USA). All other chemicals were of analytical grade. Fresh solutions of C, EC, EGC and EGCG were prepared as a stock of 3.0 mM in double distilled water (ddH$_2$O) before use as a stock of 1 mM solution. Upon addition to reaction mixtures, in the presence of buffers mentioned and at concentrations used, all the catechins used remained in solution. The volumes of stock solution added did not lead to any appreciable change in the pH of reaction mixtures. Breast cancer cell line MDA-MB-231 was maintained in DMEM (Invitrogen, Carlsbad, CA, USA) growth media. The medium was supplemented with 10% foetal bovine serum (FBS) and 1% antimycotic antibiotic (Invitrogen, Carlsbad, CA, USA). Normal breast epithelial cells MCF10A were cultured in DMEM/F12 (Invitrogen) supplemented with 5% horse serum, 20 ng/mL EGF, 0.5 µg/mL hydrocortisone, 0.1 µg/mL cholera toxin, 10 µg/mL insulin, 100 units/mL penicillin and 100 µg/mL streptomycin. Cells were cultured in a 5% CO_2-humidified atmosphere at 37 °C. MCF10A+Cu cells are MCF10A cells cultured with additional supplementation of 25 µM $CuCl_2$ for at least 4 weeks. A 5 mg/mL stock solution of 3-(4,5-dimethylthiazol-2-yl)-2,5-diphenyltetrazolium bromide (MTT) was prepared in PBS.

4.2. Absorbance Studies

The effect of increasing concentrations of Cu(II) on absorption spectra of C, EC, EGC and EGCG was observed. The reaction mixture (3.0 mL) contained 10 mMTris-HCl (pH 7.5), 50 µM C, EC, EGC, EGCG and increasing concentrations of Cu(II). The spectra were recorded immediately after addition of all components.

4.3. Flourescence Studies

The fluorescence studies were performed on a Shimadzu spectrofluorometer RF-5310 PC (Kyoto, Japan) equipped with a plotter and a calculator. C, EC, EGC and EGCG were excited at their absorption maxima (λ_{max}) of 273 nm (approximate absorption maximum of catechins). Emission spectra were recorded in the wavelength range shown in figures.

4.4. Detection of Cu(II) Reduction

The selective sequestering agent bathocuproine was employed to detect reduction of Cu(II) to Cu(I) by recording the formation of bathocuproine-Cu(I) complex which absorbs maximally at

480 nm. The reaction mixture (3.0 mL) contained 3.0 mM Tris-HCl (pH 7.5), fixed concentrations (100 μM) of Cu(II) and Cu(I) (for positive control), bathocuproine (300 μM) and of catechins (C, EC, EGC and EGCG) (50 μM). The reaction was started by adding Cu(II) and the spectra were recorded immediately afterwards.

4.5. Detection of Superoxide Anion Generation

Superoxide (O_2^-) was detected by the reduction of nitroblue tetrazolium (NBT) essentially as described by Nakayama *et al.* [28]. A typical assay mixture contained 50 mM sodium phosphate buffer (pH 7.5), 33 μM NBT, 100 μM EDTA and 0.06% triton X-100 in a total volume of 3.0 mL. The reaction was started by the addition of catechins (C/EC/EGC/EGCG). After mixing, absorbance was recorded at 560 nm at different time intervals, against a blank, which did not contain the compound.

4.6. Detection of Hydroxyl Radical Generation

In order to compare the hydroxyl radical production by increasing concentrations of C, EC, EGC and EGCG in the presence of 100 μM Cu(II), the method of Quinlan and Gutteridge [30] was followed. Calf thymus DNA (200 μg) was used as a substrate and the malondialdehyde generated from deoxyribose radicals was assayed by recording the absorbance at 532 nm.

4.7. Degradation of Calf Thymus DNA

Single strand specific digestion was performed as described by Wani and Hadi [44]. Reaction mixtures (0.5 mL) contained 10 mM Tris-HCl (pH 7.5), 500 μg of calf thymus DNA and varying amounts of C, EC, EGC, EGCG and cupric chloride (50 μM). All solutions were sterilized before use. Incubation was performed at 37 °C for one hour. The assay determines the acid soluble nucleotides released from DNA as a result of enzyme digestion. Reaction mixture in a total volume of 1.0 mL contained 40 mM Tris-HCl (pH 7.5), 1 mM Magnesium Chloride, water and enzyme. The reaction mixture was incubated at 37 °C for 2 h. The reaction was stopped by adding 0.2 mL bovine serum albumin (10 mg/mL) and 1.0 mL of 14% perchloric acid (chilled). The tubes were immediately transferred to 0 °C for 45 min before centrifugation at 2500 rpm for 10 min at room temperature to remove undigested DNA and precipitated protein. Acid soluble deoxyribonucleotides were determined in the supernatant, colorimetrically, using the diphenylamine method [45]. To a 1.0 mL aliquot, 2.0 mL diphenyl reagent (freshly prepared by dissolving 1 gram of recrystallized diphenylamine in 100 mL glacial acetic acid and 2.75 mL of concentrated H_2SO_4) was added. The tubes were heated in a boiling water bath for 30 min. The intensity of blue color was read at 600 nm.

4.8. Treatment of pBR322 DNA

Reaction mixture (30 μL) contained 10 mMTris-HCl (pH 7.5), 0.5 μg of plasmid DNA and other components as indicated in legends. Incubation was performed at 37 °C for 2 h. After incubation, 10 μL of solution containing 40 mM EDTA, 0.05% bromophenol blue (tracking dye) and 50% (v/v) glycerol was added and the solution was subjected to electrophoresis in submarine 1% agarose gel. The gel was stained with ethidium bromide (0.5 μg/mL), viewed and photographed on a UV-transilluminator.

4.9. Isolation of Lymphocytes

Heparinized blood samples (2 mL) from a single, healthy, non-smoking donor was obtained by venepuncture and diluted suitably in Ca^{2+} and Mg^{2+} free PBS. Lymphocytes were isolated from blood using Histopaque 1077 (Sigma Diagnostics, St Louis, MS, USA), and the cells were finally suspended in RPMI 1640.

4.10. Viability Assessment of Lymphocytes

The lymphocytes were checked for their viability before the start and after the end of the reaction using Trypan Blue Exclusion Test by Pool-Zobel *et al.* [46]. The viability of the cells was found to be greater than 93%.

4.11. Comet Assay

Comet assay was performed under alkaline conditions essentially according to the procedure of Singh *et al.* [47] with slight modifications. Fully frosted microscopic slides precoated with 1.0% normal melting agarose at about 50 °C (dissolved in Ca^{2+} and Mg^{2+} free PBS) were used. Approximately 10,000 cells were mixed with 75 µL of 2.0% LMPA to form a cell suspension and pipetted over the first layer and covered immediately by a coverslip. The agarose layer was allowed to solidify by placing the slides on a flat tray and keeping it on ice for 10 min. The coverslips were removed and a third layer of 0.5% LMPA (75 µL) was pipetted and coverslips placed over it and kept on ice for 5 min for proper solidification of layer. The coverslips were removed and the slides were immersed in cold lysing solution containing 2.5 M NaCl, 100 mM EDTA, 10 mM Tris, pH 10, and 1% Triton X-100 added just prior to use for a minimum of 1 h at 4 °C. After lysis DNA was allowed to unwind for 30 min in alkaline electrophoretic solution consisting of 300 mM NaOH, 1 mM EDTA, pH > 13. Electrophoresis was performed at 4 °C in a field strength of 0.7 V/cm and 300 mA current. The slides were then neutralized with cold 0.4 M Tris, pH 7.5, stained with 75 µL Ethidium Bromide (20 µg/mL) and covered with a coverslip. The slides were placed in a humidified chamber to prevent drying of the gel and analyzed the same day. Slides were scored using an image analysis system (Komet 5.5, Kinetic Imaging, Liverpool, UK) attached to a Olympus (CX41) fluorescent microscope and a COHU 4910 (equipped with a 510–560 nm excitation and 590 nm barrier filters) integrated CC camera. Comets were scored at 100× magnification. Images from 50 cells (25 from each replicate slide) were analyzed. The parameter taken to assess lymphocytes DNA damage was tail length (migration of DNA from the nucleus, µm) and was automatically generated by Komet 5.5 image analysis system.

Treatment of intact lymphocytes with four catechins (C, EC, EGC and EGCG) and the subsequent Comet assay was performed essentially as described earlier by Azmi *et al.* [48]. For antioxidant study [49], the cells were preincubated with polyphenols in eppendorf tubes in a reaction volume of 1.0 mL. After the preincubation (for 30 min at 37 °C), the reaction mixture was centrifuged at 4000 rpm, the supernatant was discarded and the pelleted lymphocytes were resuspended in 100 µL of PBS (Ca^{2+} and Mg^{2+} free) and layered for further treatment with TBHP (50 µM). The incubation period was 30 min at 37 °C in dark. The other conditions remained the same as described above.

4.12. Determination of TBARS

Thiobarbituric acid reactive substance was determined according to the method of Ramanathan *et al.* [50]. A cell suspension (1×10^5/mL) was incubated with C, EC, EGC and EGCG (0–200 µM) at 37 °C for 1 h and then centrifuged at 1000 rpm. In some experiments the cells were pre-incubated with fixed concentrations of neocuproine and thiourea. The cell pellet was washed twice with phosphate buffered saline (Ca^{2+} and Mg^{2+} free) and suspended in 0.1 N NaOH. This cell suspension (1.4 mL) was further treated with 10% TCA and 0.6 M TBA (2-thiobarbituric acid) in boiling water bath for 10 min. The absorbance was read at 532 nm and converted into nmoles of TBA reactive substance using the molar extinction coefficient ($1.56 \times 9 \times 105 \ M^{-1} \cdot cm^{-1}$).

4.13. Cell Growth Inhibition Studies by MTT Assay

MDA-MB-231 cells were seeded at a density of 1×10^4 cells per well in 96-well microtiter culture plates. After overnight incubation, normal growth medium was removed and replaced with either fresh medium (untreated control) or different concentrations of respective catechins in growth medium. After the desired time of incubation (24 or 48 h), MTT solution was added to each well (0.1 mg/mL in

DMEM) and incubated further for 4 hours at 37 °C. Upon termination, the supernatant was aspirated and the MTT formazan, formed by metabolically viable cells, was dissolved in a solubilisation solution containing DMSO (100 μL) by mixing for 5 min on a gyratory shaker. The absorbance was measured at 540 nm (reference wavelength 690 nm) on an Ultra Multifunctional Microplate Reader (Bio-Rad, Hercules, CA, USA). Absorbance of control (without treatment) was considered as 100% cell survival. Each treatment had four replicate wells and the mean values were plotted.

4.14. Statistics

The statistical analysis was performed as described by Tice *et al.* [51] and is expressed as mean ± SEM/SD of three independent experiments. A student's *t*-test was used for examining statistically significant differences. Analysis of variance was performed using ANOVA. *p* values < 0.05 were considered statistically significant.

Acknowledgments: Acknowledgments: Authors are thankful to BSR, UGC, New Delhi, for providing financial assistance for the research work. We are also thankful to the Department of Biochemistry, AMU, Aligarh for providing us the necessary facilities.

Author Contributions: Author Contributions: MF, HYK, MO, AAH, NR and HA performed experiments. MF analyzed data and drafted manuscript. AA helped in the manuscript draft. SMH conceptualized the study, provided facilities and oversaw the project.

Conflicts of Interest: Conflicts of Interest: The authors declare no conflict of interest.

References

1. Adlercreutz, C.H.; Goldin, B.R.; Gorbach, S.L.; Hockerstedt, K.A.; Watanabe, S.; Hamalainen, E.K.; Markkanen, M.H.; Makela, T.H.; Wahala, K.T.; Adlercreutz, T. Soybean phytoestrogen intake and cancer risk. *J. Nutr.* **1995**, *125*, 757S–770S. [PubMed]
2. Park, O.J.; Surh, Y.J. Chemopreventive potential of epigallocatechin gallate and genistein: Evidence from epidemiological and laboratory studies. *Toxicol. Lett.* **2004**, *150*, 43–56. [CrossRef] [PubMed]
3. Barnes, S.; Peterson, G.; Grubbs, C.; Setchell, K. Potential role of dietary isoflavones in the prevention of cancer. *Adv. Exp. Med. Biol.* **1994**, *354*, 135–147. [PubMed]
4. Surh, Y.J. Cancer chemoprevention with dietary phytochemicals. *Nat. Rev. Cancer* **2003**, *3*, 768–780. [CrossRef] [PubMed]
5. Khan, H.Y.; Zubair, H.; Faisal, M.; Ullah, M.F.; Farhan, M.; Sarkar, F.H.; Ahmad, A.; Hadi, S.M. Plant polyphenol induced cell death in human cancer cells involves mobilization of intracellular copper ions and reactive oxygen species generation: A mechanism for cancer chemopreventive action. *Mol. Nutr. Food Res.* **2014**, *58*, 437–446. [CrossRef] [PubMed]
6. Yang, G.Y.; Liao, J.; Kim, K.; Yurkow, E.J.; Yang, C.S. Inhibition of growth and induction of apoptosis in human cancer cell lines by tea polyphenols. *Carcinogenesis* **1998**, *19*, 611–616. [CrossRef] [PubMed]
7. Shamim, U.; Hanif, S.; Ullah, M.F.; Azmi, A.S.; Bhat, S.H.; Hadi, S.M. Plant polyphenols mobilize nuclear copper in human peripheral lymphocytes leading to oxidatively generated DNA breakage: Implications for an anticancer mechanism. *Free Radic. Res.* **2008**, *42*, 764–772. [CrossRef] [PubMed]
8. Ullah, M.F.; Ahmad, A.; Khan, H.Y.; Zubair, H.; Sarkar, F.H.; Hadi, S.M. The prooxidant action of dietary antioxidants leading to cellular DNA breakage and anticancer effects: Implications for chemotherapeutic action against cancer. *Cell Biochem. Biophys.* **2013**, *67*, 431–438. [CrossRef] [PubMed]
9. Clement, M.V.; Hirpara, J.L.; Chawdhury, S.H.; Pervaiz, S. Chemopreventive agent resveratrol, a natural product derived from grapes, triggers cd95 signaling-dependent apoptosis in human tumor cells. *Blood* **1998**, *92*, 996–1002. [PubMed]
10. Kuo, M.L.; Huang, T.S.; Lin, J.K. Curcumin, an antioxidant and anti-tumor promoter, induces apoptosis in human leukemia cells. *Biochim. Biophys. Acta* **1996**, *1317*, 95–100. [CrossRef]
11. Chang, K.L.; Cheng, H.L.; Huang, L.W.; Hsieh, B.S.; Hu, Y.C.; Chih, T.T.; Shyu, H.W.; Su, S.J. Combined effects of terazosin and genistein on a metastatic, hormone-independent human prostate cancer cell line. *Cancer Lett.* **2009**, *276*, 14–20. [CrossRef] [PubMed]

12. Gupta, S.; Hastak, K.; Ahmad, N.; Lewin, J.S.; Mukhtar, H. Inhibition of prostate carcinogenesis in tramp mice by oral infusion of green tea polyphenols. *Proc. Natl. Acad. Sci. USA* **2001**, *98*, 10350–10355. [CrossRef] [PubMed]

13. Orsolic, N.; Knezevic, A.; Sver, L.; Terzic, S.; Hackenberger, B.K.; Basic, I. Influence of honey bee products on transplantable murine tumours. *Vet. Comp. Oncol.* **2003**, *1*, 216–226. [CrossRef] [PubMed]

14. Chen, Z.P.; Schell, J.B.; Ho, C.T.; Chen, K.Y. Green tea epigallocatechin gallate shows a pronounced growth inhibitory effect on cancerous cells but not on their normal counterparts. *Cancer Lett.* **1998**, *129*, 173–179. [CrossRef]

15. Said Ahmad, M.; Fazal, F.; Rahman, A.; Hadi, S.M.; Parish, J.H. Activities of flavonoids for the cleavage of DNA in the presence of cu(ii): Correlation with generation of active oxygen species. *Carcinogenesis* **1992**, *13*, 605–608. [CrossRef] [PubMed]

16. Gali, H.U.; Perchellet, E.M.; Klish, D.S.; Johnson, J.M.; Perchellet, J.P. Hydrolyzable tannins: Potent inhibitors of hydroperoxide production and tumor promotion in mouse skin treated with 12-*o*-tetradecanoylphorbol-13-acetate *in vivo*. *Int. J. Cancer* **1992**, *51*, 425–432. [CrossRef] [PubMed]

17. Hadi, S.M.; Asad, S.F.; Singh, S.; Ahmad, A. Putative mechanism for anticancer and apoptosis-inducing properties of plant-derived polyphenolic compounds. *IUBMB Life* **2000**, *50*, 167–171. [PubMed]

18. Hadi, S.M.; Bhat, S.H.; Azmi, A.S.; Hanif, S.; Shamim, U.; Ullah, M.F. Oxidative breakage of cellular DNA by plant polyphenols: A putative mechanism for anticancer properties. *Semin. Cancer Biol.* **2007**, *17*, 370–376. [CrossRef] [PubMed]

19. Bhat, S.H.; Azmi, A.S.; Hanif, S.; Hadi, S.M. Ascorbic acid mobilizes endogenous copper in human peripheral lymphocytes leading to oxidative DNA breakage: A putative mechanism for anticancer properties. *Int. J. Biochem. Cell Biol.* **2006**, *38*, 2074–2081. [CrossRef] [PubMed]

20. Kagawa, T.F.; Geierstanger, B.H.; Wang, A.H.; Ho, P.S. Covalent modification of guanine bases in double-stranded DNA. The 1.2-A Z-DNA structure of d(CGCGCG) in the presence of $CuCl_2$. *J. Biol. Chem.* **1991**, *266*, 20175–20184. [PubMed]

21. Ebadi, M.; Swanson, S. The status of zinc, copper, and metallothionein in cancer patients. *Prog. Clin. Biol. Res.* **1988**, *259*, 161–175. [PubMed]

22. Margalioth, E.J.; Udassin, R.; Cohen, C.; Maor, J.; Anteby, S.O.; Schenker, J.G. Serum copper level in gynecologic malignancies. *Am. J. Obstet. Gynecol.* **1987**, *157*, 93–96. [CrossRef]

23. Yoshida, D.; Ikeda, Y.; Nakazawa, S. Quantitative analysis of copper, zinc and copper/zinc ratio in selected human brain tumors. *J. Neuro-Oncol.* **1993**, *16*, 109–115. [CrossRef]

24. Ebara, M.; Fukuda, H.; Hatano, R.; Saisho, H.; Nagato, Y.; Suzuki, K.; Nakajima, K.; Yukawa, M.; Kondo, F.; Nakayama, A.; *et al.* Relationship between copper, zinc and metallothionein in hepatocellular carcinoma and its surrounding liver parenchyma. *J. Hepatol.* **2000**, *33*, 415–422. [CrossRef]

25. Zheng, L.F.; Wei, Q.Y.; Cai, Y.J.; Fang, J.G.; Zhou, B.; Yang, L.; Liu, Z.L. DNA damage induced by resveratrol and its synthetic analogues in the presence of Cu (II) ions: Mechanism and structure-activity relationship. *Free Radic. Biol. Med.* **2006**, *41*, 1807–1816. [CrossRef] [PubMed]

26. Farhan, M.; Zafar, A.; Chibber, S.; Khan, H.Y.; Arif, H.; Hadi, S.M. Mobilization of copper ions in human peripheral lymphocytes by catechins leading to oxidative DNA breakage: A structure activity study. *Arch. Biochem. Biophys.* **2015**, *580*, 31–40. [CrossRef] [PubMed]

27. Simpson, J.A.; Narita, S.; Gieseg, S.; Gebicki, S.; Gebicki, J.M.; Dean, R.T. Long-lived reactive species on free-radical-damaged proteins. *Biochem. J.* **1992**, *282*, 621–624. [CrossRef] [PubMed]

28. Nakayama, T.; Kimura, T.; Kodama, M.; Nagata, C. Generation of hydrogen peroxide and superoxide anion from active metabolites of naphthylamines and aminoazo dyes: Its possible role in carcinogenesis. *Carcinogenesis* **1983**, *4*, 765–769. [CrossRef] [PubMed]

29. Rahman, A.; Shahabuddin; Hadi, S.M.; Parish, J.H.; Ainley, K. Strand scission in DNA induced by quercetin and Cu(II): Role of Cu(I) and oxygen free radicals. *Carcinogenesis* **1989**, *10*, 1833–1839. [CrossRef] [PubMed]

30. Quinlan, G.J.; Gutteridge, J.M. Oxygen radical damage to DNA by rifamycin sv and copper ions. *Biochem. Pharmacol.* **1987**, *36*, 3629–3633. [CrossRef]

31. Smith, C.; Halliwell, B.; Aruoma, O.I. Protection by albumin against the pro-oxidant actions of phenolic dietary components. *Food Chem. Toxicol.* **1992**, *30*, 483–489. [CrossRef]

32. Dubuisson, M.L.; de Wergifosse, B.; Trouet, A.; Baguet, F.; Marchand-Brynaert, J.; Rees, J.F. Antioxidative properties of natural coelenterazine and synthetic methyl coelenterazine in rat hepatocytes subjected to *tert*-butyl hydroperoxide-induced oxidative stress. *Biochem. Pharmacol.* **2000**, *60*, 471–478. [CrossRef]

33. Suzuki, Y.; Apostolova, M.D.; Cherian, M.G. Astrocyte cultures from transgenic mice to study the role of metallothionein in cytotoxicity of *tert*-butyl hydroperoxide. *Toxicology* **2000**, *145*, 51–62. [CrossRef]

34. Ahmad, N.; Feyes, D.K.; Nieminen, A.L.; Agarwal, R.; Mukhtar, H. Green tea constituent epigallocatechin-3-gallate and induction of apoptosis and cell cycle arrest in human carcinoma cells. *J. Natl. Cancer Inst.* **1997**, *89*, 1881–1886. [CrossRef] [PubMed]

35. Chevion, M. A site-specific mechanism for free radical induced biological damage: The essential role of redox-active transition metals. *Free Radic. Biol. Med.* **1988**, *5*, 27–37. [CrossRef]

36. Zimmer, C.; Luck, G.; Fritzsche, H.; Triebel, H. DNA-copper (II) complex and the DNA conformation. *Biopolymers* **1971**, *10*, 441–463. [CrossRef] [PubMed]

37. Gupte, A.; Mumper, R.J. Elevated copper and oxidative stress in cancer cells as a target for cancer treatment. *Cancer Treat. Rev.* **2009**, *35*, 32–46. [CrossRef] [PubMed]

38. Devi, G.S.; Prasad, M.H.; Saraswathi, I.; Raghu, D.; Rao, D.N.; Reddy, P.P. Free radicals antioxidant enzymes and lipid peroxidation in different types of leukemias. *Clin. Chim. Acta* **2000**, *293*, 53–62. [CrossRef]

39. Oberley, T.D.; Oberley, L.W. Antioxidant enzyme levels in cancer. *Histol. Histopathol.* **1997**, *12*, 525–535. [PubMed]

40. Kong, Q.; Beel, J.A.; Lillehei, K.O. A threshold concept for cancer therapy. *Med. Hypotheses* **2000**, *55*, 29–35. [CrossRef] [PubMed]

41. Van Amelsvoort, J.M.; Van Hof, K.H.; Mathot, J.N.; Mulder, T.P.; Wiersma, A.; Tijburg, L.B. Plasma concentrations of individual tea catechins after a single oral dose in humans. *Xenobiotica: Fate Foreign Compd. Biol. Syst.* **2001**, *31*, 891–901. [CrossRef] [PubMed]

42. Khan, H.Y.; Zubair, H.; Ullah, M.F.; Ahmad, A.; Hadi, S.M. Oral administration of copper to rats leads to increased lymphocyte cellular DNA degradation by dietary polyphenols: Implications for a cancer preventive mechanism. *Biometals* **2011**, *24*, 1169–1178. [CrossRef] [PubMed]

43. Amin, A.R.; Wang, D.; Zhang, H.; Peng, S.; Shin, H.J.; Brandes, J.C.; Tighiouart, M.; Khuri, F.R.; Chen, Z.G.; Shin, D.M. Enhanced anti-tumor activity by the combination of the natural compounds (−)-epigallocatechin-3-gallate and luteolin: Potential role of p53. *J. Biol. Chem.* **2010**, *285*, 34557–34565. [CrossRef] [PubMed]

44. Wani, A.A.; Hadi, S.M. Partial purification and properties of an endonuclease from germinating pea seeds specific for single-stranded DNA. *Arch. Biochem. Biophys.* **1979**, *196*, 138–146. [CrossRef]

45. Schneider, W.C. Determination of nucleic acids in tissues by pentose analysis. *Methods Enzymol* **1957**, *3*, 880–884.

46. Pool-Zobel, B.L.; Guigas, C.; Klein, R.; Neudecker, C.; Renner, H.W.; Schmezer, P. Assessment of genotoxic effects by lindane. *Food Chem. Toxicol.* **1993**, *31*, 271–283. [CrossRef]

47. Singh, N.P.; McCoy, M.T.; Tice, R.R.; Schneider, E.L. A simple technique for quantitation of low levels of DNA damage in individual cells. *Exp. Cell Res.* **1988**, *175*, 184–191. [CrossRef]

48. Azmi, A.S.; Bhat, S.H.; Hadi, S.M. Resveratrol-Cu(II) induced DNA breakage in human peripheral lymphocytes: Implications for anticancer properties. *FEBS Lett.* **2005**, *579*, 3131–3135. [CrossRef] [PubMed]

49. Kanupriya; Dipti, P.; Sharma, S.K.; Sairam, M.; Ilavazhagan, G.; Sawhney, R.C.; Banerjee, P.K. Flavonoids protect u-937 macrophages against *tert*-butylhydroperoxide induced oxidative injury. *Food Chem. Toxicol.* **2006**, *44*, 1024–1030. [CrossRef] [PubMed]

50. Ramanathan, R.; Das, N.P.; Tan, C.H. Effects of gamma-linolenic acid, flavonoids, and vitamins on cytotoxicity and lipid peroxidation. *Free Radic. Biol. Med.* **1994**, *16*, 43–48. [CrossRef]

51. Tice, R.R.; Strauss, G.H. The single cell gel electrophoresis/comet assay: A potential tool for detecting radiation-induced DNA damage in humans. *Stem Cells* **1995**, *13* (Suppl. 1), 207–214. [PubMed]

![toxins logo] *toxins*

MDPI

Article

Tenuifolide B from *Cinnamomum tenuifolium* Stem Selectively Inhibits Proliferation of Oral Cancer Cells via Apoptosis, ROS Generation, Mitochondrial Depolarization, and DNA Damage

Chung-Yi Chen [1,†], Ching-Yu Yen [2,3,†], Hui-Ru Wang [4], Hui-Ping Yang [5], Jen-Yang Tang [6,7,8], Hurng-Wern Huang [9], Shih-Hsien Hsu [5,*] and Hsueh-Wei Chang [4,10,11,12,*]

[1] Department of Nutrition and Health Sciences, School of Medical and Health Sciences, Fooyin University, Kaohsiung 83102, Taiwan; XX377@fy.edu.tw

[2] Department of Oral and Maxillofacial Surgery Chi-Mei Medical Center, Tainan 71004, Taiwan; ycysmc@gmail.com

[3] School of Dentistry, Taipei Medical University, Taipei 11031, Taiwan

[4] Department of Biomedical Science and Environmental Biology, Kaohsiung Medical University, Kaohsiung 80708, Taiwan; whr0319@gmail.com

[5] Graduate Institute of Medicine, College of Medicine, Kaohsiung Medical University, Kaohsiung 80708, Taiwan; kayyang1950@yahoo.com.tw

[6] Department of Radiation Oncology, Faculty of Medicine, College of Medicine, Kaohsiung Medical University, Kaohsiung 80708, Taiwan; reyata@kmu.edu.tw

[7] Department of Radiation Oncology, Kaohsiung Medical University Hospital, Kaohsiung 80708, Taiwan

[8] Department of Radiation Oncology, Kaohsiung Municipal Ta-Tung Hospital, Kaohsiung 80145, Taiwan

[9] Institute of Biomedical Science, National Sun Yat-Sen University, Kaohsiung 80424, Taiwan; sting@mail.nsysu.edu.tw

[10] Institute of Medical Science and Technology, National Sun Yat-sen University, Kaohsiung 80424, Taiwan

[11] Cancer Center, Translational Research Center, Kaohsiung Medical University Hospital, Kaohsiung Medical University, Kaohsiung 80708, Taiwan

[12] Center for Research Resources and Development of Kaohsiung Medical University, Kaohsiung 80708, Taiwan

* Correspondence: jackhsu@kmu.edu.tw (S.-H.H.); changhw@kmu.edu.tw (H.-W.C.);
Tel.: +886-7-312-1101 (ext. 6353) (S.-H.H.); +886-7-312-1101 (ext. 2691) (H.-W.C.);
Fax: +886-7-312-0701 (S.-H.H.); +886-7-312-5339 (H.-W.C.)

† These authors contributed equally to this work.

Academic Editor: Carmela Fimognari
Received: 15 January 2016; Accepted: 19 October 2016; Published: 5 November 2016

Abstract: The development of drugs that selectively kill oral cancer cells but are less harmful to normal cells still provide several challenges. In this study, the antioral cancer effects of tenuifolide B (TFB), extracted from the stem of the plant *Cinnamomum tenuifolium* are evaluated in terms of their effects on cancer cell viability, cell cycle analysis, apoptosis, oxidative stress, and DNA damage. Cell viability of oral cancer cells (Ca9-22 and CAL 27) was found to be significantly inhibited by TFB in a dose-responsive manner in terms of ATP assay, yielding IC_{50} = 4.67 and 7.05 μM (24 h), but are less lethal to normal oral cells (HGF-1). Dose-responsive increases in subG1 populations as well as the intensities of flow cytometry-based annexin V/propidium iodide (PI) analysis and pancaspase activity suggested that apoptosis was inducible by TFB in these two types of oral cancer cells. Pretreatment with the apoptosis inhibitor (Z-VAD-FMK) reduced the annexin V intensity of these two TFB-treated oral cancer cells, suggesting that TFB induced apoptosis-mediated cell death to oral cancer cells. Cleaved-poly (ADP-ribose) polymerase (PARP) and cleaved-caspases 3, 8, and 9 were upregulated in these two TFB-treated oral cancer cells over time but less harmful for normal oral HGF-1 cells. Dose-responsive and time-dependent increases in reactive oxygen species (ROS) and decreases in mitochondrial membrane potential (MitoMP) in these two TFB-treated oral cancer cells suggest that TFB may generate oxidative stress as measured by flow cytometry. *N*-acetylcysteine (NAC) pretreatment reduced the TFB-induced ROS generation and further validated that ROS was relevant to TFB-induced cell death. Both flow cytometry and Western blotting demonstrated that the DNA double strand marker γH2AX dose-responsively increased in TFB-treated Ca9-22 cells and time-dependently increased in two TFB-treated oral cancer cells. Taken together, we infer that TFB can selectively inhibit cell proliferation of oral cancer cells through apoptosis, ROS generation, mitochondrial membrane depolarization, and DNA damage.

Keywords: selective killing; oral cancer; natural product; apoptosis; ROS; DNA damage

1. Introduction

Oral cancer is the sixth most common type of cancer globally [1] and is especially prevalent in areas that feature a high frequency of betel nut, alcohol, and cigarette consumption [2–4]. Oral cancer is likely ignored by patients in early stages and is commonly detected at a later stage. Late detection, combined with poor chemotherapy outcomes, leads to high morbidity and mortality rates of oral cancer [5]. While several drugs have proven effective at killing cancer cells, they are also toxic to normal tissue cells, and the need for selective antioral cancer drugs remains urgent.

A growing number of studies have reported that natural products are potent resources for anticancer drug discovery [6–11]. Many bioactive extracts and isolated compounds have been extracted from the bark of *Cinnamomum* of the Formosan Lauraceous family (*C. zeylanicum* and *C. cassia*) [12], leaves (*C. wilsonii* [13], *C. kotoense* [14–17], *C. subavenium* [18]), stems (*C. subavenium* [19,20]), and heartwood and roots (*C. osmophloeum* [21]). These findings indicate the antiproliferative effect of *Cinnamomum* plants for several types of cancer, such as that of the colon [12,13,17], lung [14,16], liver [15,21], breast [17], prostate [18,20], melanoma [19], and bladder [20]. However, the selective killing effect of *Cinnamomum* plants on oral cancer cells remains undetermined.

To try to discover new compounds from other *Cinnamomum* plants, we extracted material from *C. tenuifolium* Sugimoto form. nervosum (Meissn.) Hara [22], an evergreen form of the Lauraceae plant family grown on Orchid Island of Taiwan. Methanol extracts were used to identify a new benzodioxocinone, benzodioxocinone (2,3-dihydro-6,6-dimethylbenzo-[b][1,5]dioxocin-4(6*H*)-one), from the leaves of *C. tenuifolium* [23]. The benzodioxocinone showed mild levels of cytotoxicity for human oral cancer (OC2), with an IC_{50} value of 107.7 µM after 24 h of treatment.

Alternatively, we previously used the stems of *C. tenuifolium* [22] to identify several novel compounds, including tenuifolide A, isotenuifolide A, tenuifolide B (TFB), secotenuifolide A, and tenuifolin, along with some known compounds. Secotenuifolide A was found to provide the best antiproliferative effect against two human prostate cancer cells (DU145 and LNCaP) with IC_{50} values < 7 µM after 24 h of treatment. For TFB (3-(1-methoxyeicosyl)-5-methylene-5*H*-furan-2-one), its IC_{50} values were 246 and 22.2 µM for DU145 and LNCaP cancer cells after 24 h of treatment [22]. However, the biological effect of TFB against the oral cancer cells was not addressed as yet.

The current study first evaluates the possible selectively antiproliferactive effect and the mechanism of *C. tenuifolium* stem-derived TFB on oral cancer cells by analyzing cell viability, cell cycle progression, apoptosis, reactive oxygen species (ROS) induction, mitochondrial depolarization, and DNA damage.

2. Results

2.1. Cell Viability and ATP Cellular Content

ATP content has been widely used to measure cell viability [24,25]. Figure 1 shows the ATP assay of cell viability after 24 h of treatment with TFB (0, 5, 10, and 15 µM). The viability of TFB-treated oral cancer cells (Ca9-22 and CAL 27) decreased dose-responsively ($p < 0.001$). In contrast, the normal oral cells (HGF-1) maintained a cell viability of about 100%.

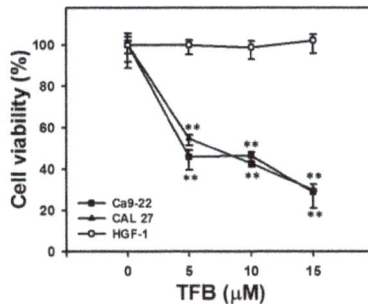

Figure 1. Tenuifolide B (TFB) induced a significant decrease in ATP-based cell viability in oral cancer cells (Ca9-22 and CAL 27) but not in normal oral cells (HGF-1). Cells were treated with 0, 5, 10, and 15 μM TFB for 24 h. Data: mean ± SD ($n = 4$). ** $p < 0.001$ compared to the control.

2.2. Cell Cycle Progression

To examine whether the cell cycle was affected by TFB, the cell cycle progression was examined. Figure 2A,B show dose-responsive pattern changes of the cell cycle progression of TFB-treated Ca9-22 and CAL 27 cells, respectively. The subG1 population in TFB-treated Ca9-22 and CAL 27 cells increased in a dose-responsive manner after 24 h of THB treatment (Figure 2C,D) ($p < 0.001$).

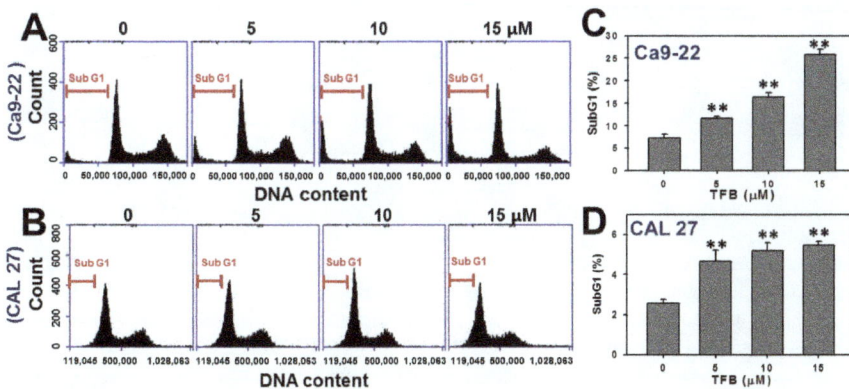

Figure 2. TFB induced an increase in the subG1 population in oral cancer Ca9-22 and CAL 27 cells. (**A,B**) Representative dose responses of cell phase profiles in TFB-treated Ca9-22 and CAL 27 cells using flow cytometry. Cells were treated with 0, 5, 10, and 15 μM TFB for 24 h. (**C,D**) Quantification analysis results for subG1 population in (**A,B**). Data: mean ± SD ($n = 3$). ** $p < 0.001$ compared to the control.

2.3. Annexin V-Based Apoptosis

To validate the role of apoptosis in the increase in the subG1 population in TFB-treated Ca9-22 and CAL 27 cells, the annexin V/propidium iodide (PI) staining method was used. Figure 3A,B respectively show the patterns of dose response changes of annexin V/PI staining profiles of TFB-treated Ca9-22 and CAL 27 cells. By calculating the percentages of annexin V positive (%), the apoptosis level (Figure 3C,D) show a significant increase in a dose-responsive manner in TFB-treated Ca9-22 and CAL 27 cells ($p < 0.001$). When the Ca9-22 and CAL 27 cells were pretreated with apoptosis inhibitor Z-VAD-FMK, apoptosis induced by different doses of TFB was decreased.

Figure 3E,F show the patterns of time course changes of annexin V/PI staining profiles of TFB-treated Ca9-22 and CAL 27 cells. The degree of annexin V/PI staining in TFB-treated Ca9-22 and CAL 27 cells increased in a time-dependent manner ($p < 0.001$) (Figure 3G,H).

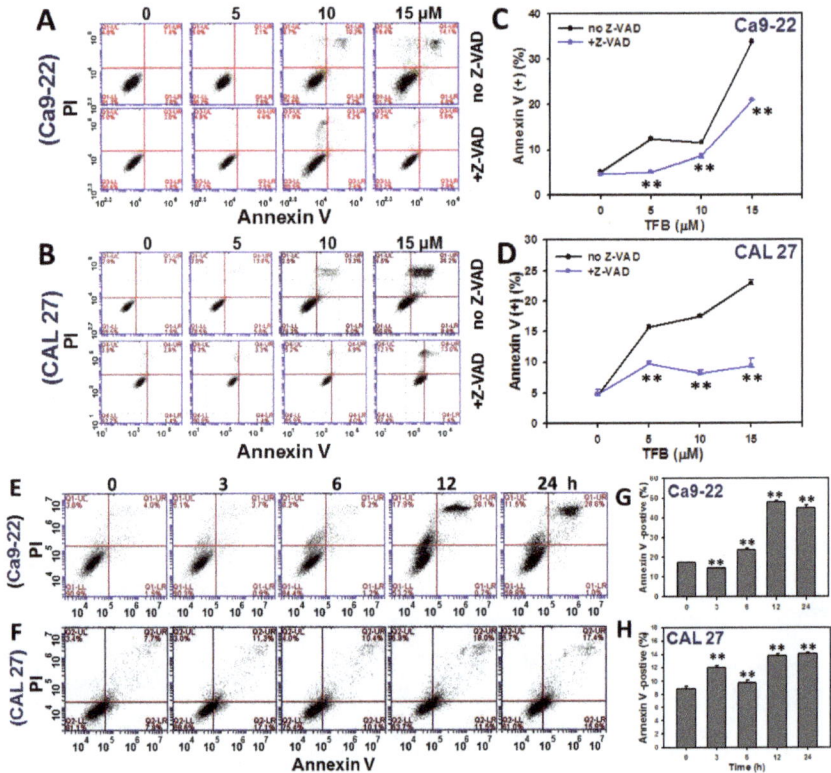

Figure 3. TFB induced annexin V-based apoptosis in oral cancer Ca9-22 and CAL 27 cells. (**A,B**) Representative flow cytometry-based dose response of apoptosis profiles of annexin V/PI double staining for TFB-treated Ca9-22 and CAL 27 cells. Cells were treated with 0, 5, 10, and 15 µM TFB for 24 h with or without 0.1 mM Z-VAD-FMK treatment for 2 h. (**C,D**) Quantification analysis results for apoptosis positive (%) in (**A,B**). Data: mean ± SD ($n = 3$). ** $p < 0.001$, comparing the Z-VAD-FMK to the non-Z-VAD-FMK for each TFB dose. (**E,F**) Representative time course of annexin V-based apoptosis profile in TFB-treated Ca9-22 and CAL 27 cells using flow cytometry. Cells were treated with 15 µM TFB for 3, 6, 12, and 24 h. (**G,H**) Quantification analysis results for annexin V-based apoptosis in (E,F). The regions of annexin V (+) including annexin V (+)/PI (+) and annexin V (+)/PI (−) were analyzed. Data: mean ± SD ($n = 3$). ** $p < 0.001$ compared to the control.

2.4. Caspases-Based Apoptosis

To validate the role of apoptosis in the increase in annexin V intensity in TFB-treated Ca9-22 and CAL 27 cells, the pancaspase activity assay was used. Figure 4A,B respectively show the dose-responsive pattern changes of pancaspase intensity profiles of TFB (0, 5, 10, and 15 µM)-treated Ca9-22 and CAL 27 cells. For the pancaspase positive (%), Figure 4C,D respectively show a significant increase in the apoptosis expression in TFB-treated Ca9-22 and CAL 27 cells at higher doses ($p < 0.001$).

Figure 4E,F show the patterns of time course changes of pancaspase intensity profiles of TFB-treated Ca9-22 and CAL 27 cells. The pancaspase intensities in TFB-treated Ca9-22 and CAL

27 cells increased in a time-dependent manner (Figure 4G,H) ($p < 0.001$). Moreover, Figure 4I shows the protein expressions of apoptosis signaling proteins, such as cleaved-poly (ADP-ribose) polymerase (PARP) and cleaved-caspases 3 and 8 gradually increased over 3–24 h and cleaved-caspase 9 was detected at 24 h in TFB-treated Ca9-22 cells. In TFB-treated CAL 27 cells, the protein expressions of cleaved-PARP and cleaved-caspases 3, 8, and 9 gradually increased from 3 to 6 h, moderately increased at 12 h, and declined at 24 h. In contrast, these apoptosis signaling proteins in TFB-treated HGF-1 were weak.

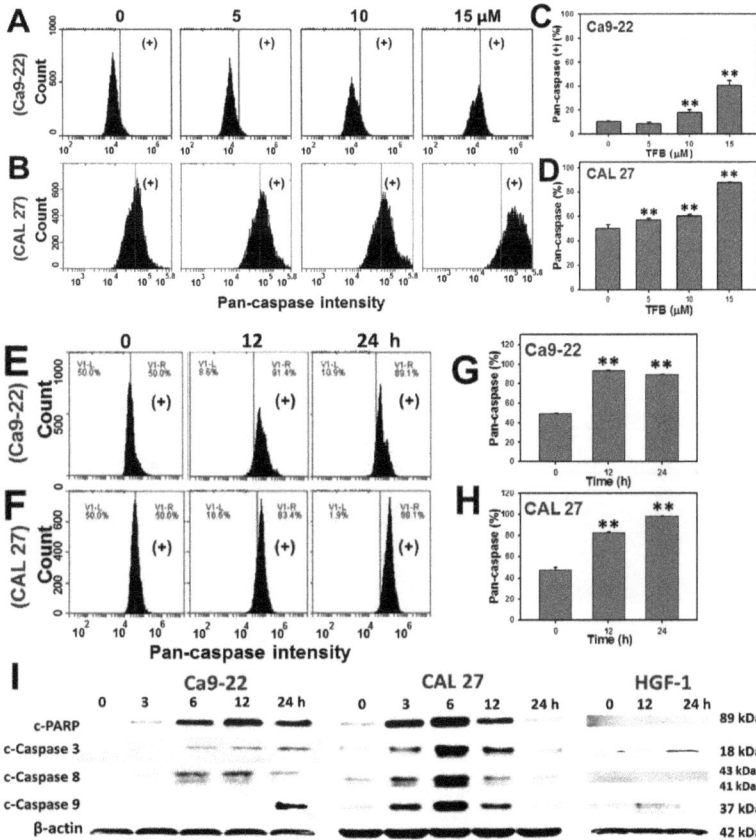

Figure 4. TFB induced caspases-based apoptosis in oral cancer Ca9-22 and CAL 27 cells. (**A,B**) Representative flow cytometry-based dose response of apoptosis profiles of pancaspase staining for TFB-treated Ca9-22 and CAL 27 cells. Cells were treated with 0, 5, 10, and 15 μM TFB for 24 h and subsequently stained with TF2-VAD-FMK. (**C,D**) Quantification analysis results for pancaspase fluorescent intensity positive (%) in (**A,B**). (**E,F**) Representative time course of pancaspase-based apoptosis profile in TFB-treated Ca9-22 and CAL 27 cells using flow cytometry. Cells were treated with 15 μM TFB for 12 and 24 h. (**G,H**) Quantification analysis results for pancaspase-based apoptosis in (**E,F**). The regions of pancaspase (+) were analyzed. Data: mean ± SD ($n = 3$). ** $p < 0.001$ compared to the control. (**I**) The apoptosis-related protein expressions in TFB-treated Ca9-22, CAL 27, and HGF-1 cells. Ca9-22 and CAL 27 cells were treated with 15 μM TFB for 3, 6, 12, and 24 h. HGF-1 cells were treated with 15 μM TFB for 12 and 24 h. Apoptosis signaling proteins, such as cleavage forms of PARP, procaspase 3, procaspase 8, and procaspase 9 were detected using Western blotting. The β-actin was used as an internal control.

2.5. ROS

To determine why TFB may inhibit cancer cell proliferation and induce apoptosis, the cellular ROS level was examined. Figure 5A,B respectively show the dose-responsive pattern changes of ROS change profiles of TFB (0, 5, 10, and 15 µM)-treated Ca9-22 and CAL 27 cells. By calculating the percentages of DCFH-DA fluorescence-positive intensity and adjusting with control, the relative ROS level was increased in a dose-responsive manner in TFB-treated Ca9-22 and CAL 27 cells ($p < 0.001$) (Figure 5C,D). When cells were pretreated by N-acetylcysteine (NAC) in Ca9-22 and CAL 27 cells, ROS generation induced by different doses of TFB was decreased.

Figure 5E–G show the time course pattern changes of ROS intensity profiles of TFB-treated Ca9-22, CAL 27, and HGF-1 cells, respectively. The ROS intensities in TFB-treated Ca9-22 and CAL 27 cells dramatically increased in a time-dependent manner (Figure 5H,I) ($p < 0.001$). In contrast, ROS intensities in HGF-1 cells increased only slightly after 24 h of THB treatment (Figure 5J).

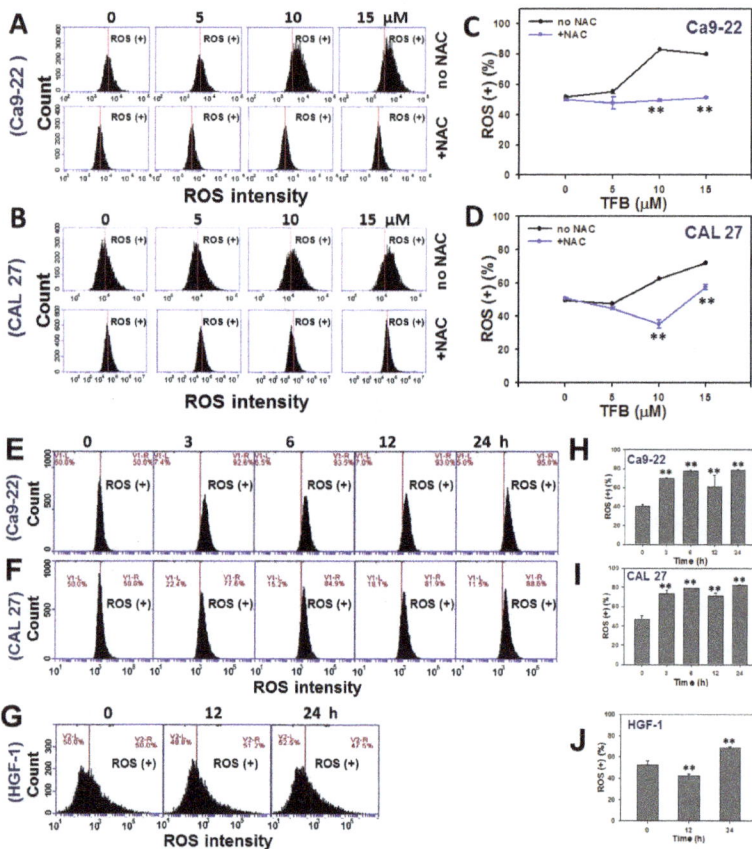

Figure 5. TFB induced ROS generation in oral cancer Ca9-22 and CAL 27 cells. (**A,B**) Representative dose responses of ROS profiles for TFB-treated cells using flow cytometry. Cells were treated with 0, 5, 10, and 15 µM of TFB for 24 h with or without 2 mM NAC pretreatment for 1 h. (**C,D**) Quantification analysis of ROS intensity for DCFD-A positivity (%). (**E–G**) Representative time course of ROS profile in TFB-treated Ca9-22, CAL 27, and HGF-1 cells using flow cytometry. Cells were treated with 15 µM TFB for indicated times. (**H–J**) Quantification analysis results for ROS intensity for ROS positivity (%) in (**E–G**). Data: mean ± SD ($n = 3$). ** $p < 0.001$ compared to the control.

2.6. Mitochondrial Membrane Potentials (MitoMP)

DiOC$_2$(3)-based MitoMP detection assay was performed to evaluate the impact of TFB (0, 5, 10, and 15 µM)-induced ROS generation. Figure 6A,B show the MitoMP profiles for TFB-treated oral cancer Ca9-22 and CAL 27 cells, respectively, after a 24 h of treatment. By calculating the percentages of DiOC$_2$(3)-negative in Figure 6A,B and comparing with the control, it was found that the MitoMP-negative (%) was gradually increased in TFB-treated Ca9-22 and CAL 27 cells in a dose-responsive manner ($p < 0.001$) (Figure 6C,D). Therefore, the MitoMP level of Ca9-22 and CAL 27 cells was significantly decreased after TFB treatment.

Figure 6E,F show the time course pattern changes of MitoMP intensity profiles of TFB-treated Ca9-22 and CAL 27 cells. The MitoMP-negative intensities in TFB-treated Ca9-22 and CAL 27 cells increased in a time-dependent manner (Figure 6G,H) ($p < 0.001$), suggesting that MitoMP levels of Ca9-22 and CAL 27 cells decreased after TFB treatment.

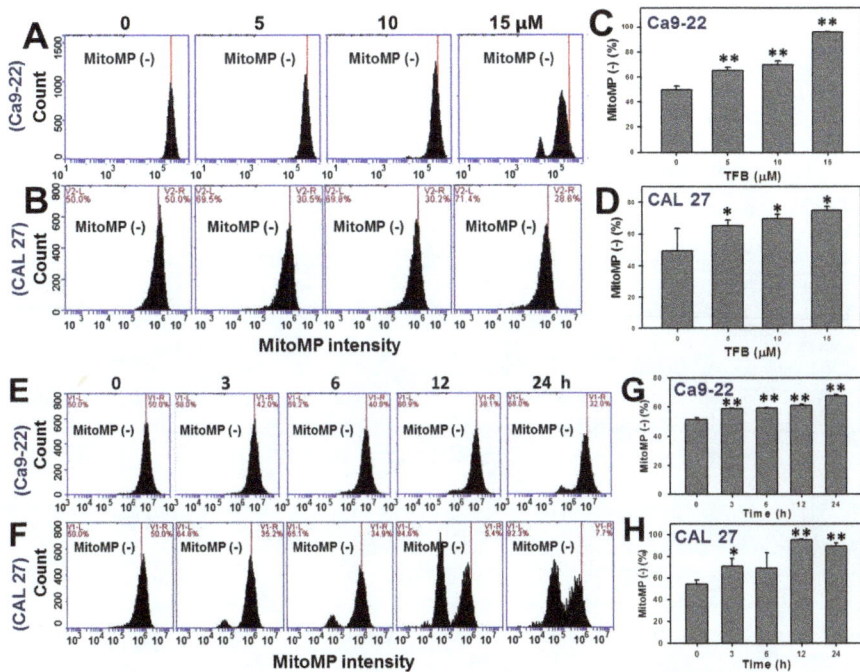

Figure 6. TFB decreased mitochondrial membrane potential (MitoMP) in oral cancer Ca9-22 and CAL 27 cells. (**A,B**) Representative dose response of MitoMP profiles for TFB-treated Ca9-22 and CAL 27 cells using flow cytometry. Cells were treated with 0, 5, 10, and 15 µM of TFB for 24 h. (**C,D**) Quantification analysis of MitoMP-negative (%) intensity in (A,B). (**E,F**) Representative time course of MitoMP profile in TFB-treated Ca9-22 and CAL 27 cells using flow cytometry. Cells were treated with 15 µM TFB for 3, 6, 12, and 24 h. (**G,H**) MitoMP-negative (%) intensity in (E,F). The regions of MitoMP-negative (%) were analyzed. Data: mean ± SD ($n = 3$). * $p < 0.05$; ** $p < 0.001$ compared to the control.

2.7. γH2AX Expression

To examine the role of DNA damage in TFB-induced antiproliferation of Ca9-22 oral cancer cells, the expression of DNA double strand break marker γH2AX was analyzed via both flow cytometry and Western blotting. Figure 7A shows that the flow cytometry-based γH2AX/PI staining profiles of TFB-treated Ca9-22 cells after 24 h of treatment. Figure 7B shows the γH2AX-positive intensity of TFB (0, 5, 10, and 15 µM)-treated Ca9-22 cells increased in a dose-responsive manner ($p < 0.001$). Moreover,

Figure 7C shows that the γH2AX expression by Western blotting of TFB-treated Ca9-22 cells after 24 h of treatment with indicated doses were dramatically increased at higher doses.

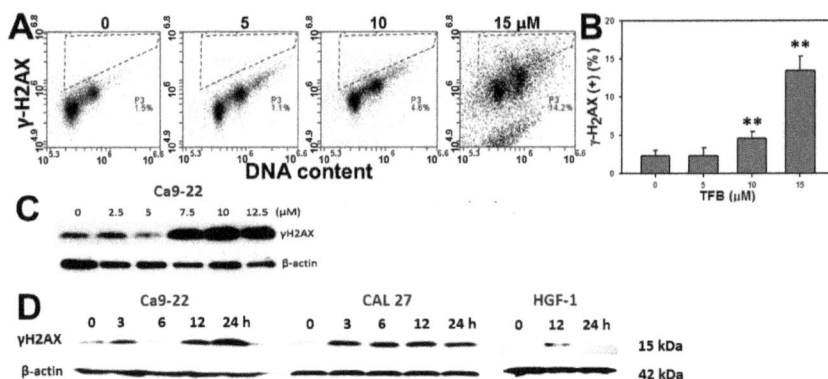

Figure 7. Treatment with TFB induced γH2AX expressions in oral cancer Ca9-22 and CAL 27 cells. (**A**) γH2AX expression of TFB-treated Ca9-22 cells was detected by flow cytometry. Cells were treated with 0, 5, 10, and 15 μM TFB for 24 h. (**B**) Quantification analysis of γH2AX positive (%) intensity of flow cytometry in (**A**). Data: mean ± SD (*n* = 3). ** *p* < 0.001 compared to the control. (**C**) Dose response of γH2AX expression of TFB-treated Ca9-22 cells was detected by Western blotting. The β-actin was used as an internal control. (**D**) Time course of γH2AX expression of TFB-treated Ca9-22, CAL 27, and HGF-1 cells was detected by Western blotting. Oral cancer cells were treated with 15 μM TFB for 0, 3, 6, 12, and 24 h. HGF-1 cells were treated with 15 μM TFB for 12 and 24 h.

For the time course experiments, Figure 7D shows that γH2AX expression of TFB-treated Ca9-22, CAL 27, and HGF-1 cells were increased in a time-dependent manner. The TFB-induced γH2AX levels were dramatically induced in both oral cancer cells (Ca9-22 and CAL 27). In contrast, TFB-induced γH2AX levels increased only slightly at 12 h in oral normal HGF-1 cells.

3. Discussion

TFB was previously found to be anti-atherosclerogenic in humans [26], but their possible anticancer effect regarding oral cancer remained unclear. In general, *Cinnamomum* plants generally have antiproliferative effects. For example, the cytotoxicity (IC$_{50}$) is known for several *Cinnamomum* plants. Cinnamaldehyde from *C. zeylanicum* and *C. cassia* barks is effective against colon cancer (HT29) = 19.7 μM at 72 h [12], (3*R*,9*S*)-megastigman-5-ene-3,9-diol 3-*O*-β-D-glucopyranoside from *C. wilsonii* leaves against colon cancer SW-480 cells = 12 μM at 48 h [13], isoobtusilactone A from *C. kotoense* leaves against human hepatoma Hep G2 cells = 37.5 μM at 18 h [15], cinnakotolactone from *C. kotoense* leaves against human colorectal cancer HT29 and breast MCF-7 cells = 25.8 and 24.4 μM at 72 h [17], subamone from *C. subavenium* leaves against prostate cancer LNCaP cells = 7.01 μM at 24 h [18], subamolide B from *C. subavenium* stems against melanoma A375 cells = 17.59 μM at 48 h [19], and subamolide A from *C. subavenium* stems against human prostate cancer PC3 cells = 10.1 μM at 72 h [20]. It has to be noted that some of the IC$_{50}$ values were determined after 72 h of treatment.

Based on ATP content assays, the current study found that the IC$_{50}$ values of TFB were 4.67 and 7.05 μM after 24 h of treatment in oral cancer cells (Ca9-22 and CAL 27), respectively. In general, the sensitivity of TFB to oral cancer cells (Ca9-22 and CAL 27) was higher than that of other *Cinnamomum* plants to other types of cancer cells. Moreover, TFB was less cytotoxic to human prostate cancer DU145 and LNCaP cells [22]. These results suggest that TFB may have a cancer cell type-specific antiproliferation effect.

Following 24 h of treatment, the cytotoxicity (IC_{50}) of taxol in human prostate cancer DU145 and LNCaP cells is 4.84 and 6.32 µM, respectively [22]. The IC_{50} of cisplatin in oral cancer Ca9-22 cells is 10.2 µM (data not shown). Therefore, our developed TFB (IC_{50} = 4.67 and 7.05 µM) has similar sensitivity to these clinical drugs in oral cancer cells (Ca9-22 and CAL 27). Moreover, we found that TFB is less harmful to normal oral HGF-1 cells (Figure 1), suggesting that TFB selectively kills oral cancer cells and may prevent side effects of oral cancer therapy. Similarly, *Cinnamomum* stem bark extract has been reported to selectively kill other types of cancer cells. In *Cinnamomum burmannii* Blume stem bark extract after 24 h of treatment, the IC_{50} values of nasopharyngeal carcinoma cells (HK1 and C666-1) were 108.32 and 224.32 µg/mL, respectively, whereas the IC_{50} of immortalized human skin keratinocyte HaCaT cells was 320.29 µg/mL [27]. However, our study only tested one normal oral cell line, and further study is needed to confirm these findings, including more normal oral cell lines to further show the lack of possible side-effects of TFB.

Secotenuifolide A, also isolated from the same material of this study (*C. tenuifolium* stems), has been reported to inhibit cell proliferation, increase the subG1 population, induce apoptosis and ROS generation, and decrease mitochondrial membrane potential in human prostate cancer cells, DU145 [22]. Secotenuifolide A also exhibited a release of cytochrome c from mitochondria and the activation of caspase-9/caspase-3 [22]. Similarly, the TFB from *C. tenuifolium* stems showed the same effect of oxidative stress (ROS induction and MitoMP depletion) on oral cancer Ca9-22 and CAL 27 cells (Figures 5 and 6). In contrast, the TFB-induced ROS generation in HGF-1 cells was only slightly induced. These results suggest that TFB exhibited selective ROS induction in oral cancer cells (Ca9-22 and CAL 27), but less induction in HGF-1 cells. NAC pretreatment experiments (Figure 5A,B) validated that ROS was relevant to TFB-induced cell death because the ROS generation in the two types of oral cancer cells was reduced by NAC pretreatment.

Annexin V and pancaspase results (Figures 3 and 4) support that TFB is apoptosis-inducible in oral cancer cells. Moreover, the TFB-induced apoptosis was reduced by Z-VAD-FMK pretreatment (Figure 3A,B), suggesting that apoptosis was involved in selective killing by TFB. However, the role of apoptosis signaling in TFB-induced apoptosis was not addressed specifically. Caspases 8 and 9 involved in intrinsic and extrinsic apoptotic pathways, respectively. Both converge in activating the executioner caspases 3 and 7 [28]. PARP is also involved in apoptosis [29,30]. To address the role of apoptosis signaling, we found that both TFB-treated oral cancer cells (Ca9-22 and CAL 27) induced activation of PARP and caspases 3, 8, and 9 by cleavage. However, these oral cancer cells displayed a differential expression of these TFB-induced apoptosis proteins. For example, cleaved-caspase 8 was mainly or early induced during 3–12 h in TFB-treated Ca9-22 cells, but cleaved-caspase 9 induction showed a later response at 24 h (Figure 4I). For CAL 27 cells, cleaved-PARP and cleaved-caspases 3, 8 and 9 were upregulated early after 3 h. These caspases peaked at 6 h, gradually declining by 24 h. Similarly, other drug-induced apoptosis also showed a similar tendency for cleaved-caspase expression. For example, cleaved-caspase 3 increased at 6–12 h and declined at 24 h in 0.2 µM staurosporine-treated human endothelial cornea cells [31]. After treatment of 0.2 µM staurosporine, cleaved-caspase 3 was also increased at 2–8 h and declined at 12–24 h for human cervical cancer HeLa cells. Cleaved-caspase 3 increased at 0.5–1 h and declined at 2–24 h in cervical cancer C-33A cells [32].

Mounting evidence demonstrated that ROS generation and mitochondrial membrane depolarization may lead to DNA damage [33–38] and apoptosis [38–41] in drug-treated cancer cells. Accordingly, TFB also showed a correlation between oxidative stress and DNA damage in its antiproliferative and apoptotic effects of two oral cancer cells (Ca9-22 and CAL 27) in our study.

4. Conclusions

TFB treatment induces apoptosis, ROS generation, mitochondrial depolarization, and DNA damage, which ultimately results in the antiproliferation of oral cancer Ca9-22 cells. This study also shows that TFB selectively kills the two oral cancer cell lines tested here and opts for its application in anti-oral cancer therapies.

5. Materials and Methods

5.1. Drug Information and Oral Cancer and Normal Cell Lines

The TFB ($C_{26}H_{46}O_3$; MW: 406.3447; [3-(1-methoxyeicosyl)-5-methylene-5H-furan-2-one]) was purified from methanol extracts of *C. tenuifolium* stem as previously described [22]. It was dissolved in dimethyl sulfoxide (DMSO) for drug treatments. The oral cancer cell lines Ca9-22 [42] and CAL 27 [43] were incubated in a mixed medium with Dulbecco's Modified Eagle Medium (DMEM) and F12 (Gibco, Grand Island, NY, USA) (3:2), 10% fetal bovine serum, antibiotics (penicillin and streptomycin), and others under a humidified atmosphere with 5% CO_2 at 37 °C. The normal oral cells (human gingival fibroblasts, HGF-1) were incubated in a DMEM medium with a similar supplement with 1 mM pyruvate as described above.

5.2. Measurement of Cell Viability—Cellular ATP Content

Cellular ATP level was determined by using the ATP-lite Luminescence ATP Detection Assay System (PerkinElmer Life Sciences, Boston, MA, USA) according to the manufacturer's instructions with a slight modification [44]. Briefly, Ca9-22 cells were plated at 4000 cells/well in 96-well plates. After seeding overnight, cells were treated with vehicle control (DMSO) or with TFB at indicated concentrations (5, 10, and 15 μM) for 24 h. After removing the medium solution, 100 μL of serum-free medium and 50 μL of a mammalian cell lysis solution was added per well of a microplate with orbital shaking at 100 rpm for 5 min. Then, 100 μL of the cell lysates/well was transferred to white 96-well plates and reacted with 50 μL of substrate solution (D-Luciferin and luciferase) under orbital shaking at 100 rpm for 5 min and then left to stand in darkness for a further 10 min. Finally, the luminescence was assayed using a microplate luminometer (CentroPRO LB 962, Berthold, ND, USA).

5.3. Measurement of Cell Cycle Progression

The cellular DNA was stained with PI as previously described [45]. Cells were plated at 3×10^5 cells/2 mL cell culture medium on a 6-well plate. Briefly, cells were added with vehicle (DMSO only) or TFB. After collection for 70% ethanol fixation overnight, the centrifuged cell pellets were resuspended in 1 mL of PBS containing 50 μg/mL PI for 15 min at room temperature in darkness. Subsequently, these samples were examined using a FACSCalibur flow cytometer (Becton-Dickinson, Mansfield, MA, USA) (excitation: 488 nm and emission: 617 nm) and BD Accuri C6 software.

5.4. Measurement of Apoptosis by Annexin V Staining

Annexin V (Strong Biotech Corporation, Taipei, Taiwan)/PI (Sigma, St. Louis, MO, USA) double-staining for apoptosis analysis was performed as previously described [46]. Cells were plated at 3×10^5 cells/2 mL cell culture medium on a 6-well plate with or without 0.1 mM Z-VAD-FMK pretreatments for 2 h (Selleckchem.com; Houston, TX, USA). Briefly, cells were added with vehicle or TFB. The cells were then resuspended in the binding buffer containing 5 μg/mL of annexin V-fluorescein isothiocyanate and 50 μg/mL of PI and examined with a BD Accuri C6 flow cytometer (Becton-Dickinson, Mansfield, MA, USA) (excitation: 488 nm and emission: 525 nm and 617 nm for FITC and PI, respectively) and BD Accuri C6 software.

5.5. Measurement of Apoptosis by Caspase Activity

Apoptosis was also measured by caspase activation [47]. The generic caspase activity assay kit (Fluorometric-Green; ab112130) (Abcam, Cambridge, UK) was used to detect the activity of caspase-1, -3, -4, -5, -6, -7, -8, and -9 as described [42]. Briefly, cells were seeded as 3×10^5 cells per well in 6-well plates with a 2 mL medium for overnight. The cells were then treated with vehicle or TFB. Subsequently, cells were incubated at 37 °C, 5% CO_2 for 2 h with 2 μL of 500X TF2-VAD-FMK. After PBS washing, cells were resuspended in 0.5 mL of an assay buffer for immediate flow cytometry measurement (BD Accuri™ C6; Becton-Dickinson).

The apoptosis signaling expressions were further measured via Western blotting. 30 µg protein lysates were resolved in 10% SDS-PAGE. After electrotransferring, the nonspecific bindings of PDVF membranes (Pall Corporation, Port Washington, NY, USA) were blocked with 5% non-fat milk in Tris-buffered saline with Tween-20 and incubated with primary antibodies (the cleaved caspase-8 (Asp391) (18C8) rabbit mAb and the rabbit mAb in the apoptosis antibody sampler kit (cleaved PARP (Asp214) (D64E10) XP®; cleaved caspase-3 (Asp175) (5A1E); cleaved caspase-9 (Asp330) (D2D4) from Cell Signalling Technology, Inc., Danvers, MA, USA, β-actin (#GTX629630, GeneTex Inc.)) under 1:10000 dilution as well as their matched secondary antibody. The WesternBright™ ECL HRP substrate (#K-12045-D50, Advansta, Menlo Park, CA, USA) was chosen for signal amplification.

5.6. Measurement of Intracellular ROS

2′,7′-Dichlorodihydrofluorescein diacetate (DCFH-DA) (Sigma Chemical Co., St. Louis, MO, USA) was used to detect intracellular ROS as previously described [33]. Cells were plated at 3×10^5 cells/2 mL cell culture medium on a 6 cm dish. Briefly, cells were added with vehicle or TFB with or without 2 mM NAC pretreatment for 1 h (Sigma; St. Louis, MO, USA). After the collection and PBS washing, cells were treated with 0.1 µM DCFH-DA in serum-free medium for 30 min at 37 °C in darkness. Cells were resuspended in PBS after centrifugation and examined with a BD Accuri C6 flow cytometer (excitation: 488 nm and emission: 525 nm) and BD Accuri C6 software.

5.7. Measurement of MitoMP

A MitoProbe™ DiOC$_2$(3) assay kit (Invitrogen, San Diego, CA, USA) was used to detect mitochondrial membrane potential (MitoMP) as described previously [34]. Cells were plated at 3×10^5 cells/2 mL cell culture medium on a 6-well plate. Briefly, cells were added with vehicle or with TFB. The TFB-treated cells were washed in 1 mL of PBS/well, provided with 1 mL of medium/well, loaded with 10 µL of 10 µM DiOC$_2$(3), and left to stand at 37 °C in 5% CO$_2$ for 20–30 min. After harvesting and washing, cells were resuspended in PBS and examined immediately using a FACSCalibur flow cytometer (excitation: 488 nm and emission: 525 nm) and BD Accuri C6 software.

5.8. Measurement of DNA Damage by γH2AX Expression

DNA double strand breaks were detected by both flow cytometry [48] and Western blotting [35] as described previously. For flow cytometry, TFB-treated cells were fixed, washed, and incubated at 4 °C for 1 h in 2 µg/mL of p-Histone H2AX (Ser 139) (γH2AX) monoclonal antibody (sc-101696; Santa Cruz Biotechnology, Santa Cruz, CA, USA). After washing, cells were suspended for 1 h in a secondary antibody (Jackson Laboratory, Bar Harbor, ME, USA) for 30 min at room temperature. Finally, the cells were resuspended in 20 µg/mL of PI for flow cytometry analysis (BD Accuri™ C6; Becton-Dickinson).

For Western blotting of γH2AX expression, 30 µg protein lysates were resolved in 10% SDS-PAGE. Except when p-Histone H2AX (Santa Cruz Biotechnology) was chosen for the primary antibody, procedures were the same as those employing Western blotting for apoptosis proteins, mentioned above.

5.9. Statistical Analysis

All data are shown as mean ± SD. The significant differences between test and control were analyzed with a Student *t*-test.

Acknowledgments: This work was partly supported by funds of the Ministry of Science and Technology (NSC 101-2320-B-037-049, MOST 103-2320-B-037-008, MOST 103-2320-B-242-001, and MOST 104-2320-B-037-013-MY3), the Chimei-KMU jointed project (105CM-KMU-06), the National Sun Yat-sen University-KMU Joint Research Project (#NSYSUKMU 105-P0322), the Kaohsiung Municipal Ta-Tung Hospital (kmtth-104-003), the Kaohsiung Medical

University "Aim for the Top Universities Grant, grant No. KMU-TP104PR02", and the Health and Welfare Surcharge of Tobacco Products, Ministry of Health and Welfare, Taiwan, Republic of China (MOHW105-TDU-B-212-134007). We also thank Hans-Uwe Dahms for the help of English editing.

Author Contributions: Hui-Ru Wang and Hui-Ping Yang carried out the experiments. Ching-Yu Yen, Jen-Yang Tang, Hurng-Wern Huang, and Shih-Hsien Hsu analyzed the data. Chung-Yi Chen, Ching-Yu Yen, and Hsueh-Wei Chang conceived and designed the study. Chung-Yi Chen, Shih-Hsien Hsu, and Hsueh-Wei Chang wrote and revised the manuscript.

Conflicts of Interest: The authors declare no conflict of interest.

References

1. Warnakulasuriya, S. Global epidemiology of oral and oropharyngeal cancer. *Oral Oncol.* **2009**, *45*, 309–316. [CrossRef] [PubMed]
2. Ko, Y.C.; Huang, Y.L.; Lee, C.H.; Chen, M.J.; Lin, L.M.; Tsai, C.C. Betel quid chewing, cigarette smoking and alcohol consumption related to oral cancer in Taiwan. *J. Oral Pathol. Med.* **1995**, *24*, 450–453. [CrossRef] [PubMed]
3. Lee, C.H.; Ko, Y.C.; Huang, H.L.; Chao, Y.Y.; Tsai, C.C.; Shieh, T.Y.; Lin, L.M. The precancer risk of betel quid chewing, tobacco use and alcohol consumption in oral leukoplakia and oral submucous fibrosis in southern Taiwan. *Br. J. Cancer* **2003**, *88*, 366–372. [CrossRef] [PubMed]
4. Chiang, S.L.; Lee, C.P.; Chang, J.G.; Lee, C.H.; Yeh, K.T.; Tsai, Y.S.; Chen, M.K.; Chen, C.H.; Ko, Y.C. Combined effects of differentiation factor 15 and substance use of alcohol, betel quid and cigarette on risk of head and neck cancer. *Head Neck Oncol.* **2013**, *5*, 23.
5. Myoung, H.; Hong, S.P.; Yun, P.Y.; Lee, J.H.; Kim, M.J. Anti-cancer effect of genistein in oral squamous cell carcinoma with respect to angiogenesis and in vitro invasion. *Cancer Sci.* **2003**, *94*, 215–220. [CrossRef] [PubMed]
6. Chen, B.H.; Chang, H.W.; Huang, H.M.; Chong, I.W.; Chen, J.S.; Chen, C.Y.; Wang, H.M. (−)-Anonaine induces DNA damage and inhibits growth and migration of human lung carcinoma h1299 cells. *J. Agric. Food Chem.* **2011**, *59*, 2284–2290. [CrossRef] [PubMed]
7. Wang, H.M.; Cheng, K.C.; Lin, C.J.; Hsu, S.W.; Fang, W.C.; Hsu, T.F.; Chiu, C.C.; Chang, H.W.; Hsu, C.H.; Lee, A.Y. Obtusilactone A and (−)-sesamin induce apoptosis in human lung cancer cells by inhibiting mitochondrial Lon protease and activating DNA damage checkpoints. *Cancer Sci.* **2010**, *101*, 2612–2620. [CrossRef] [PubMed]
8. Hseu, Y.C.; Wu, C.R.; Chang, H.W.; Kumar, K.J.; Lin, M.K.; Chen, C.S.; Cho, H.J.; Huang, C.Y.; Lee, H.Z.; Hsieh, W.T.; et al. Inhibitory effects of Physalis angulata on tumor metastasis and angiogenesis. *J. Ethnopharmacol.* **2011**, *135*, 762–771. [CrossRef] [PubMed]
9. Russo, M.; Spagnuolo, C.; Tedesco, I.; Russo, G.L. Phytochemicals in cancer prevention and therapy: Truth or dare? *Toxins* **2010**, *2*, 517–551. [CrossRef] [PubMed]
10. Melchini, A.; Traka, M.H. Biological profile of erucin: A new promising anticancer agent from cruciferous vegetables. *Toxins* **2010**, *2*, 593–612. [CrossRef] [PubMed]
11. Yadav, V.R.; Prasad, S.; Sung, B.; Kannappan, R.; Aggarwal, B.B. Targeting inflammatory pathways by triterpenoids for prevention and treatment of cancer. *Toxins* **2010**, *2*, 2428–2466. [CrossRef] [PubMed]
12. Cabello, C.M.; Bair, W.B., 3rd; Lamore, S.D.; Ley, S.; Bause, A.S.; Azimian, S.; Wondrak, G.T. The cinnamon-derived Michael acceptor cinnamic aldehyde impairs melanoma cell proliferation, invasiveness, and tumor growth. *Free Radic. Biol. Med.* **2009**, *46*, 220–231. [CrossRef] [PubMed]
13. Shu, P.; Wei, X.; Xue, Y.; Li, W.; Zhang, J.; Xiang, M.; Zhang, M.; Luo, Z.; Li, Y.; Yao, G.; et al. Wilsonols A–L, megastigmane sesquiterpenoids from the leaves of *Cinnamomum wilsonii*. *J. Nat. Prod.* **2013**, *76*, 1303–1312. [CrossRef] [PubMed]
14. Chen, C.Y.; Chen, C.H.; Lo, Y.C.; Wu, B.N.; Wang, H.M.; Lo, W.L.; Yen, C.M.; Lin, R.J. Anticancer activity of isoobtusilactone A from *Cinnamomum kotoense*: Involvement of apoptosis, cell-cycle dysregulation, mitochondria regulation, and reactive oxygen species. *J. Nat. Prod.* **2008**, *71*, 933–940. [CrossRef] [PubMed]
15. Chen, C.Y.; Liu, T.Z.; Chen, C.H.; Wu, C.C.; Cheng, J.T.; Yiin, S.J.; Shih, M.K.; Wu, M.J.; Chern, C.L. Isoobtusilactone A-induced apoptosis in human hepatoma Hep G2 cells is mediated via increased NADPH oxidase-derived reactive oxygen species (ROS) production and the mitochondria-associated apoptotic mechanisms. *Food Chem. Toxicol.* **2007**, *45*, 1268–1276. [CrossRef] [PubMed]

16. Chen, C.Y.; Hsu, Y.L.; Chen, Y.Y.; Hung, J.Y.; Huang, M.S.; Kuo, P.L. Isokotomolide A, a new butanolide extracted from the leaves of *Cinnamomum kotoense*, arrests cell cycle progression and induces apoptosis through the induction of p53/p21 and the initiation of mitochondrial system in human non-small cell lung cancer A549 cells. *Eur. J. Pharmacol.* **2007**, *574*, 94–102. [PubMed]

17. Yang, S.S.; Hou, W.C.; Huang, L.W.; Lee, T.H. A new gamma-lactone from the leaves of *Cinnamomum kotoense*. *Nat. Prod. Res.* **2006**, *20*, 1246–1250. [CrossRef] [PubMed]

18. Lin, R.J.; Lo, W.L.; Wang, Y.D.; Chen, C.Y. A novel cytotoxic monoterpenoid from the leaves of *Cinnamomum subavenium*. *Nat. Prod. Res.* **2008**, *22*, 1055–1059. [CrossRef] [PubMed]

19. Yang, S.Y.; Wang, H.M.; Wu, T.W.; Chen, Y.J.; Shieh, J.J.; Lin, J.H.; Ho, T.F.; Luo, R.J.; Chen, C.Y.; Chang, C.C. Subamolide B isolated from medicinal plant *Cinnamomum subavenium* induces cytotoxicity in human cutaneous squamous cell carcinoma cells through mitochondrial and CHOP-dependent cell death pathways. *Evid Based Complement. Altern. Med.* **2013**, *2013*, 630415. [CrossRef] [PubMed]

20. Liu, C.H.; Chen, C.Y.; Huang, A.M.; Li, J.H. Subamolide A, a component isolated from *Cinnamomum subavenium*, induces apoptosis mediated by mitochondria-dependent, p53 and ERK1/2 pathways in human urothelial carcinoma cell line NTUB1. *J. Ethnopharmacol.* **2011**, *137*, 503–511. [CrossRef] [PubMed]

21. Chen, T.H.; Huang, Y.H.; Lin, J.J.; Liau, B.C.; Wang, S.Y.; Wu, Y.C.; Jong, T.T. Cytotoxic lignan esters from *Cinnamomum osmophloeum*. *Planta Med.* **2010**, *76*, 613–619. [CrossRef] [PubMed]

22. Lin, R.J.; Cheng, M.J.; Huang, J.C.; Lo, W.L.; Yeh, Y.T.; Yen, C.M.; Lu, C.M.; Chen, C.Y. Cytotoxic compounds from the stems of *Cinnamomum tenuifolium*. *J. Nat. Prod.* **2009**, *72*, 1816–1824. [CrossRef] [PubMed]

23. Chen, H.L.; Kuo, S.Y.; Li, Y.P.; Kang, Y.F.; Yeh, Y.T.; Huang, J.C.; Chen, C.Y. A new benzodioxocinone from the leaves of *Cinnamomum tenuifolium*. *Nat. Prod. Res.* **2012**, *26*, 1881–1886. [CrossRef] [PubMed]

24. Tam, K.F.; Ng, T.Y.; Liu, S.S.; Tsang, P.C.; Kwong, P.W.; Ngan, H.Y. Potential application of the ATP cell viability assay in the measurement of intrinsic radiosensitivity in cervical cancer. *Gynecol. Oncol.* **2005**, *96*, 765–770. [CrossRef] [PubMed]

25. Lu, X.; Errington, J.; Chen, V.J.; Curtin, N.J.; Boddy, A.V.; Newell, D.R. Cellular ATP depletion by LY309887 as a predictor of growth inhibition in human tumor cell lines. *Clin. Cancer Res.* **2000**, *6*, 271–277. [PubMed]

26. Dong, H.P.; Wu, H.M.; Chen, S.J.; Chen, C.Y. The effect of butanolides from *Cinnamomum tenuifolium* on platelet aggregation. *Molecules* **2013**, *18*, 11836–11841. [CrossRef] [PubMed]

27. Daker, M.; Lin, V.Y.; Akowuah, G.A.; Yam, M.F.; Ahmad, M. Inhibitory effects of *Cinnamomum burmannii* Blume stem bark extract and trans-cinnamaldehyde on nasopharyngeal carcinoma cells; synergism with cisplatin. *Exp. Ther. Med.* **2013**, *5*, 1701–1709. [CrossRef] [PubMed]

28. Tait, S.W.; Green, D.R. Mitochondria and cell death: Outer membrane permeabilization and beyond. *Nat. Rev. Mol. Cell Biol.* **2010**, *11*, 621–632. [CrossRef] [PubMed]

29. Chang, Y.T.; Huang, C.Y.; Li, K.T.; Li, R.N.; Liaw, C.C.; Wu, S.H.; Liu, J.R.; Sheu, J.H.; Chang, H.W. Sinuleptolide inhibits proliferation of oral cancer Ca9–22 cells involving apoptosis, oxidative stress, and DNA damage. *Arch. Oral Biol.* **2016**, *66*, 147–154. [CrossRef] [PubMed]

30. Yen, Y.H.; Farooqi, A.A.; Li, K.T.; Butt, G.; Tang, J.Y.; Wu, C.Y.; Cheng, Y.B.; Hou, M.F.; Chang, H.W. Methanolic extracts of *Solieria robusta* inhibits proliferation of oral cancer Ca9–22 cells via apoptosis and oxidative stress. *Molecules* **2014**, *19*, 18721–18732. [CrossRef] [PubMed]

31. Thuret, G.; Chiquet, C.; Herrag, S.; Dumollard, J.M.; Boudard, D.; Bednarz, J.; Campos, L.; Gain, P. Mechanisms of staurosporine induced apoptosis in a human corneal endothelial cell line. *Br. J. Ophthalmol.* **2003**, *87*, 346–352. [CrossRef] [PubMed]

32. Nicolier, M.; Decrion-Barthod, A.Z.; Launay, S.; Pretet, J.L.; Mougin, C. Spatiotemporal activation of caspase-dependent and -independent pathways in staurosporine-induced apoptosis of p53wt and p53mt human cervical carcinoma cells. *Biol. Cell* **2009**, *101*, 455–467. [CrossRef] [PubMed]

33. Yeh, C.C.; Yang, J.I.; Lee, J.C.; Tseng, C.N.; Chan, Y.C.; Hseu, Y.C.; Tang, J.Y.; Chuang, L.Y.; Huang, H.W.; Chang, F.R.; et al. Anti-proliferative effect of methanolic extract of *Gracilaria tenuistipitata* on oral cancer cells involves apoptosis, DNA damage, and oxidative stress. *BMC Complement. Altern. Med.* **2012**, *12*, 142. [CrossRef] [PubMed]

34. Yen, C.Y.; Chiu, C.C.; Haung, R.W.; Yeh, C.C.; Huang, K.J.; Chang, K.F.; Hseu, Y.C.; Chang, F.R.; Chang, H.W.; Wu, Y.C. Antiproliferative effects of goniothalamin on Ca9–22 oral cancer cells through apoptosis, DNA damage and ROS induction. *Mutat. Res.* **2012**, *747*, 253–258. [CrossRef] [PubMed]

35. Chiu, C.C.; Haung, J.W.; Chang, F.R.; Huang, K.J.; Huang, H.M.; Huang, H.W.; Chou, C.K.; Wu, Y.C.; Chang, H.W. Golden berry-derived 4beta-hydroxywithanolide E for selectively killing oral cancer cells by generating ROS, DNA damage, and apoptotic pathways. *PLoS ONE* **2013**, *8*, e64739. [CrossRef] [PubMed]

36. Guo, J.; Zhao, W.; Hao, W.; Ren, G.; Lu, J.; Chen, X. Cucurbitacin B induces DNA damage, G2/M phase arrest, and apoptosis mediated by reactive oxygen species (ROS) in leukemia K562 cells. *Anticancer Agents Med. Chem.* **2014**, *14*, 1146–1153. [CrossRef] [PubMed]

37. Ni, C.H.; Yu, C.S.; Lu, H.F.; Yang, J.S.; Huang, H.Y.; Chen, P.Y.; Wu, S.H.; Ip, S.W.; Chiang, S.Y.; Lin, J.G.; et al. Chrysophanol-induced cell death (necrosis) in human lung cancer A549 cells is mediated through increasing reactive oxygen species and decreasing the level of mitochondrial membrane potential. *Environ. Toxicol.* **2014**, *29*, 740–749. [CrossRef] [PubMed]

38. Hseu, Y.C.; Lee, M.S.; Wu, C.R.; Cho, H.J.; Lin, K.Y.; Lai, G.H.; Wang, S.Y.; Kuo, Y.H.; Kumar, K.J.; Yang, H.L. The chalcone flavokawain B induces G2/M cell-cycle arrest and apoptosis in human oral carcinoma HSC-3 cells through the intracellular ROS generation and downregulation of the Akt/p38 MAPK signaling pathway. *J. Agric. Food Chem.* **2012**, *60*, 2385–2397. [CrossRef] [PubMed]

39. Huang, F.J.; Hsuuw, Y.D.; Chan, W.H. Characterization of apoptosis induced by emodin and related regulatory mechanisms in human neuroblastoma cells. *Int. J. Mol. Sci.* **2013**, *14*, 20139–20156. [CrossRef] [PubMed]

40. Shih, H.C.; El-Shazly, M.; Juan, Y.S.; Chang, C.Y.; Su, J.H.; Chen, Y.C.; Shih, S.P.; Chen, H.M.; Wu, Y.C.; Lu, M.C. Cracking the cytotoxicity code: Apoptotic induction of 10-acetylirciformonin B is mediated through ROS generation and mitochondrial dysfunction. *Mar. Drugs* **2014**, *12*, 3072–3090. [CrossRef] [PubMed]

41. Thangam, R.; Senthilkumar, D.; Suresh, V.; Sathuvan, M.; Sivasubramanian, S.; Pazhanichamy, K.; Gorlagunta, P.K.; Kannan, S.; Gunasekaran, P.; Rengasamy, R.; et al. Induction of ROS-dependent mitochondria-mediated intrinsic apoptosis in MDA-MB-231 cells by glycoprotein from *Codium decorticatum*. *J. Agric. Food Chem.* **2014**, *62*, 3410–3421. [CrossRef] [PubMed]

42. Yeh, C.C.; Tseng, C.N.; Yang, J.I.; Huang, H.W.; Fang, Y.; Tang, J.Y.; Chang, F.R.; Chang, H.W. Antiproliferation and induction of apoptosis in Ca9–22 oral cancer cells by ethanolic extract of *Gracilaria tenuistipitata*. *Molecules* **2012**, *17*, 10916–10927. [CrossRef] [PubMed]

43. Jiang, L.; Ji, N.; Zhou, Y.; Li, J.; Liu, X.; Wang, Z.; Chen, Q.; Zeng, X. CAL 27 is an oral adenosquamous carcinoma cell line. *Oral Oncol.* **2009**, *45*, e204–e207. [CrossRef] [PubMed]

44. Wei, J.; Stebbins, J.L.; Kitada, S.; Dash, R.; Zhai, D.; Placzek, W.J.; Wu, B.; Rega, M.F.; Zhang, Z.; Barile, E.; et al. An optically pure apogossypolone derivative as potent pan-active inhibitor of anti-apoptotic bcl-2 family proteins. *Front. Oncol.* **2011**, *1*, 28. [CrossRef] [PubMed]

45. Chiu, C.C.; Chang, H.W.; Chuang, D.W.; Chang, F.R.; Chang, Y.C.; Cheng, Y.S.; Tsai, M.T.; Chen, W.Y.; Lee, S.S.; Wang, C.K.; et al. Fern plant-derived protoapigenone leads to DNA damage, apoptosis, and G(2)/m arrest in lung cancer cell line H1299. *DNA Cell Biol.* **2009**, *28*, 501–506. [CrossRef] [PubMed]

46. Chiu, C.C.; Liu, P.L.; Huang, K.J.; Wang, H.M.; Chang, K.F.; Chou, C.K.; Chang, F.R.; Chong, I.W.; Fang, K.; Chen, J.S.; et al. Goniothalamin inhibits growth of human lung cancer cells through DNA damage, apoptosis, and reduced migration ability. *J. Agric. Food Chem.* **2011**, *59*, 4288–4293. [CrossRef] [PubMed]

47. Kaufmann, S.H.; Lee, S.H.; Meng, X.W.; Loegering, D.A.; Kottke, T.J.; Henzing, A.J.; Ruchaud, S.; Samejima, K.; Earnshaw, W.C. Apoptosis-associated caspase activation assays. *Methods* **2008**, *44*, 262–272. [CrossRef] [PubMed]

48. Yen, C.Y.; Hou, M.F.; Yang, Z.W.; Tang, J.Y.; Li, K.T.; Huang, H.W.; Huang, Y.H.; Lee, S.Y.; Fu, T.F.; Hsieh, C.Y.; et al. Concentration effects of grape seed extracts in anti-oral cancer cells involving differential apoptosis, oxidative stress, and DNA damage. *BMC Complement. Altern. Med.* **2015**, *15*, 94. [CrossRef] [PubMed]

toxins

MDPI

Article

Chemical Characterization and *in Vitro* Cytotoxicity on Squamous Cell Carcinoma Cells of *Carica Papaya* Leaf Extracts

Thao T. Nguyen [1], Marie-Odile Parat [1], Mark P. Hodson [1,2], Jenny Pan [1], Paul N. Shaw [1] and Amitha K. Hewavitharana [1,*]

[1] School of Pharmacy, The University of Queensland, Brisbane, QLD 4072, Australia;
t.nguyen65@uq.edu.au (T.T.N.); m.parat@pharmacy.uq.edu.au (M.-O.P.); m.hodson1@uq.edu.au (M.P.H.);
jenny.pan@uqconnect.edu.au (J.P.); n.shaw@pharmacy.uq.edu.au (P.N.S.)
[2] Metabolomics Australia, Australian Institute for Bioengineering and Nanotechnology,
The University of Queensland, Brisbane, QLD 4072, Australia
* Correspondence: a.hewavitharana@pharmacy.uq.edu.au; Tel.: +61-7-334-61898; Fax: +61-7-334-61999

Academic Editor: Carmela Fimognari
Received: 26 October 2015; Accepted: 18 December 2015; Published: 24 December 2015

Abstract: In traditional medicine, *Carica papaya* leaf has been used for a wide range of therapeutic applications including skin diseases and cancer. In this study, we investigated the *in vitro* cytotoxicity of aqueous and ethanolic extracts of *Carica papaya* leaves on the human oral squamous cell carcinoma SCC25 cell line in parallel with non-cancerous human keratinocyte HaCaT cells. Two out of four extracts showed a significantly selective effect towards the cancer cells and were found to contain high levels of phenolic and flavonoid compounds. The chromatographic and mass spectrometric profiles of the extracts obtained with Ultra High Performance Liquid Chromatography-Quadrupole Time of Flight-Mass Spectrometry were used to tentatively identify the bioactive compounds using comparative analysis. The principal compounds identified were flavonoids or flavonoid glycosides, particularly compounds from the kaempferol and quercetin families, of which several have previously been reported to possess anticancer activities. These results confirm that papaya leaf is a potential source of anticancer compounds and warrant further scientific investigation to validate the traditional use of papaya leaf to treat cancer.

Keywords: *Carica papaya*; cytotoxicity; mass spectrometry; cancer; flavonoids; chromatography

1. Introduction

A book entitled "The most wonderful tree in the world—the papaw tree (*Carica papaia*)", published some 100 years ago, contains many anecdotes relating to the cure of breast, liver or rectal cancer after "treatment" with *Carica papaya* preparations [1]. Subsequent reports have been published in various media that have detailed "the healing capabilities of an old Australian Aboriginal remedy—boiled extract of pawpaw leaves—against cancer" [2] and several other anecdotes relating "cancer cure" following consumption of various preparations of papaya plant [3–6].

Recently, we undertook a comprehensive literature review [7] and found that research providing scientific evidence for the effectiveness of *Carica papaya* in the treatment and prevention of cancer was limited. However, in contrast to the limited number of studies that have been done to evaluate the effects of papaya extracts on cancer, the abundance in *Carica papaya* of phytochemicals with reported anticancer activities, such as carotenoids (in fruits and seeds), alkaloids (in leaves), phenolics (in fruits, leaves, shoots) and glucosinolates (in seeds and fruits), suggests that there are opportunities for new research to evaluate the anticancer potential of this medicinal plant [7].

Squamous cell carcinoma (SCC) is the second most common type of skin cancer and also occurs in many other epithelia such as lips, mouth, urinary bladder, prostate, lung and vagina. Skin squamous cell carcinomas are not only more likely to metastasize but also to cause mortality, when compared with skin basal cell carcinoma [8]. Although different parts of the *Carica papaya* plant have been used as traditional medicine for the treatment of skin infections and wound healing in general, and this widespread use has been scientifically validated [9–11], no information is available on the activity of this plant on skin cancer. Furthermore, the effects of *Carica papaya* leaf extracts have previously been reported being tested on the growth of different cancer cell lines: breast, stomach, lung, pancreatic, colon, liver, ovarian, cervical, neuroblastoma, lymphoma, leukaemia and other blood cancers [12–14]; to our knowledge, no skin cancer cell lines have been tested. We hypothesized that *Carica papaya* leaf extracts exerted *in vitro* cytotoxicity on human squamous cell carcinoma. In this study, human oral squamous cell carcinoma (SCC25) cells and immortal, non-cancerous human keratinocyte cells (HaCaT) were selected for the cytotoxic studies of papaya extracts. The HaCaT cell line was selected to permit experiments to be performed in parallel with SCC25 in order to screen for candidate extracts with selective growth inhibition towards cancer cells, a highly desirable feature of potential cancer preventative and therapeutic agents. Our aim was also to preliminarily identify the bioactive compounds using liquid chromatography-quadrupole time-of-flight-mass spectrometry (LC-QToF-MS).

2. Results and Discussion

The MTT assay has been widely applied in proliferation and cytotoxicity studies to screen the chemo-preventive potential of natural products. It provides preliminary data for further *in vitro* and *in vivo* studies. The addition of organic solvents is required to solubilize the extracts from natural products in cell culture media; therefore it is prudent to investigate the effect of the solvents on the cell lines under experimentation to identify the most suitable solvent and its optimal concentration in media. This information can then be used during sample preparation for rigorous cytotoxicity studies using the MTT assay. Dimethyl sulfoxide (DMSO) has been reported to be the solvent of choice for sample preparations with a final concentration in the medium from 0.1% to 1.0% but typically data have not been reported relating to impact of such DMSO concentrations on cell viability [15–17]. In our investigation, we found that DMSO at a concentration as low as 0.05% causes significant toxicity to SCC25 and a significantly different effect was observed between the two cell lines. In contrast, ethanol (EtOH) up to a concentration of 1.0% did not significantly impact upon the viability of either cell line (Figure 1).

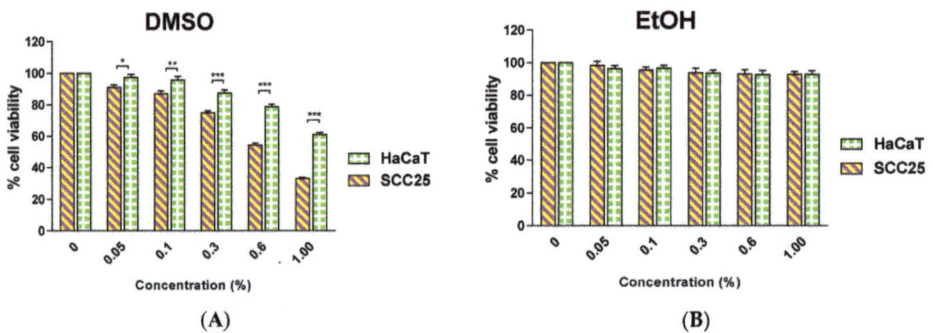

Figure 1. Effect of a 48-h incubation with Dimethyl sulfoxide (DMSO) (**A**); or Ethanol (EtOH) (**B**) on the survival of human squamous cell carcinoma (SCC25) and human keratinocyte (HaCaT) cells. Results are shown as mean $\pm$ SEM (n = 3). * $p < 0.05$, ** $p < 0.01$; *** $p < 0.001$, HaCaT *vs.* SCC25 (two-way analysis of variance (ANOVA) with Bonferroni post-tests).

Therefore, in this study, ethanol at a concentration of 0.3% was chosen as the solvent for the preparation of ethanolic extracts in media for cell experiments. To investigate the effect of the papaya leaf extracts on SCC25 and HaCaT cells, cells were treated with extracts over a range of concentrations (5–100 µg/mL) for 48 h and the percentage cell viability was analyzed. As revealed in Figure 2, all four extracts showed a significant effect on SCC25 cancer cell viability, starting at different concentrations: 25 µg/mL for serial basic ethanol (SBE), 10 µg/mL for serial acidic ethanol (SAE), and 5 µg/mL for both serial acidic water (SAW) and serial basic water (SBW) fractions. However, when the cell viability effects between SCC25 cancer cells and non-cancerous HaCaT cells were compared, the two fractions with acidic pH showed a significantly selective cytotoxicity towards the SCC25 cells with an effective range from 25–100 µg/mL for SAE and 5–20 µg/mL for SAW fractions (Figure 3).

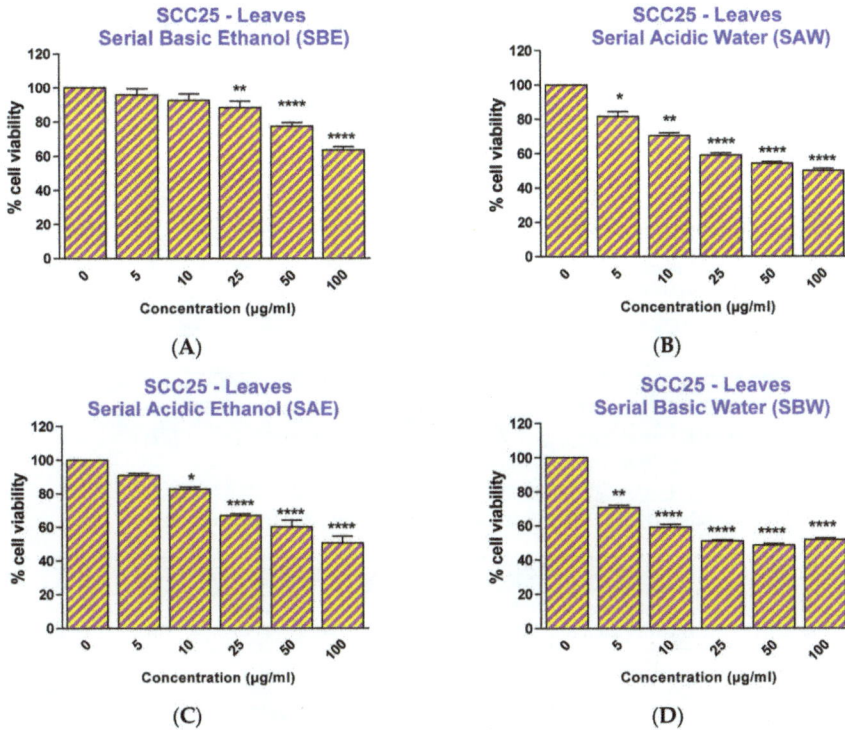

Figure 2. Effect of papaya leaf extracts (serially extracted in the listed order, with basic ethanol (**A**); acidic ethanol (**B**); acidic water (**C**); and basic water (**D**)) on the survival of SCC25 cells. Results are shown as mean $\pm$ SEM ($n = 3$). * $p < 0.05$; ** $p < 0.01$; **** $p < 0.0001$, *vs.* EtOH-treated control (one-way ANOVA with Kruskal-Wallis test).

The IC50 values clearly showed the selective effect of two acidic extracts with IC50 values for SCC25 cells smaller than thosefor HaCaT cells.This was not observed to be the case for either of the two basic extracts (Table 1). To eliminate the possibility that this selective effect might due to cancer cells being more sensitive to acidic pH than the non-cancerous cells, the pH of the medium containing extracts at the highest tested concentration (100 µg/mL) was measured. The pH of the media was unaffected by the addition of either acidic or basic extracts, likely due to the buffer capacity of the medium and the small addition of the extracts. Therefore, we conclude here that the acidic conditions provided extracts containing important compounds with selective effects on skin cancer cell viability.

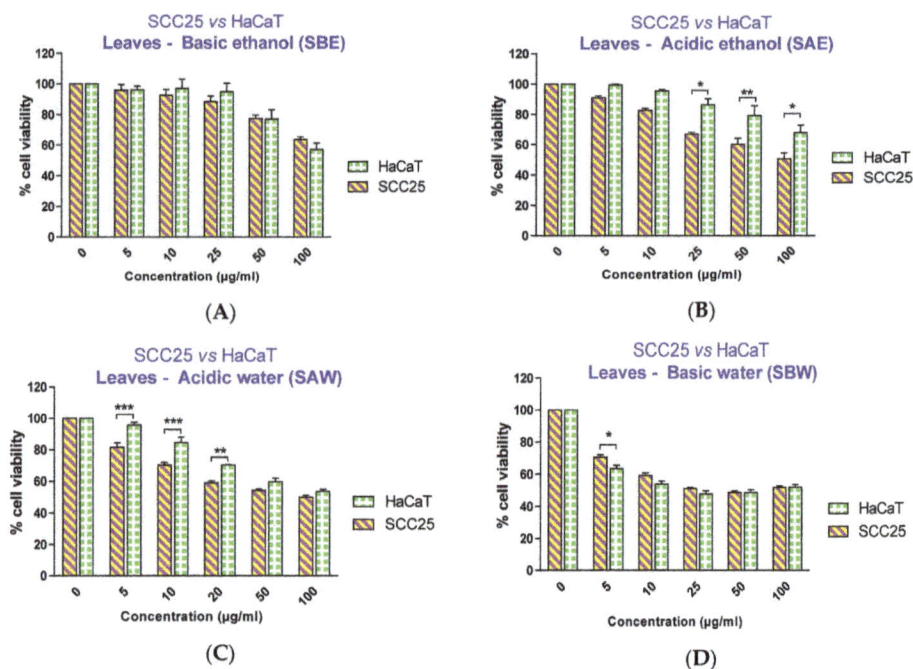

Figure 3. Effect of papaya leaf extracts (serially extracted in the listed order with basic ethanol (**A**); acidic ethanol (**B**); acidic water (**C**); and basic water (**D**)) on the survival of SCC25 and HaCaT cells. Results are shown as mean ± SEM ($n = 3$). * $p < 0.05$; ** $p < 0.01$; *** $p < 0.001$, HaCaT *vs.* SCC25 (two-way ANOVA with Bonferroni post-test).

Table 1. IC$_{50}$ values of tested extracts for SCC25 and HaCaT cells.

Cell Lines	IC$_{50}$ (µg/mL) (95% Confidence Interval)			
	Serial Basic Ethanol Extract	Serial Acidic Ethanol Extract	Serial Acidic Water Extract	Serial Basic Water Extract
SCC25	172.9 (151.6–197.3)	77.18 (62.71–94.99)	57.72 (41.99–79.35)	40.14 (27.03–59.61)
HaCaT	157.6 (119.8–207.4)	199.5 (155.4–256.2)	85.74 (71.63–102.6)	34.24 (21.47–54.60)

Interestingly, examination of the phenolic and flavonoid content of the same four fractions (Figure 4) indicated that the phenolic and flavonoid contents positively correlated with the differential effect on cell viability as shown in Figure 3. Acidic ethanolic and water fractions had much higher flavonoid content (SAE = 15.60 ± 0.07; SAW = 9.95 ± 0.05 mg quercetin equivalents (QE)/g extract) compared to the basic fractions that contained less than 2.00 mg QE/g (SBE = 0.63 ± 0.30; SBW = 1.90 ± 0.23 mg QE/g extract). The phenolic content of acidic fractions was also found to be higher than that in basic fractions but to a lesser extent than the flavonoid content (SAW = 62.98 ± 0.30; SAE = 44.26 ± 0.27 mg gallic acid equivalents (GAE)/g extract and SBW = 40.18 ± 0.16; SBE = 26.92 ± 2.53 mg GAE/g extract).

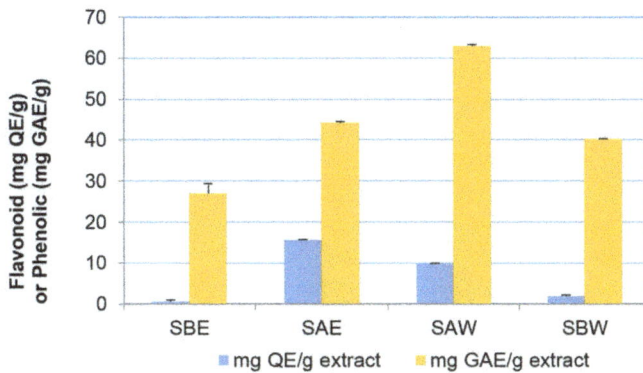

Figure 4. Total flavonoid and phenolic content in the extracts. Results are shown as mean ± SEM (*n* = 3).

The apparent positive correlation between the selective effects on the SCC25 cancer cells and the content of flavonoids and phenolic compounds initiated further comparative analysis of the chemical constituents to identify the compounds that are present exclusively or at higher concentrations in the acidic extracts compared to basic extracts. The chromatographic data were extracted to features by Molecular Feature Extractor algorithms and then aligned by Mass Profiler Professional software (Version 12.1, Agilent Technologies, Santa Clara, CA, USA, 2012). In positive ionization mode, a total of 432 and 191 features were detected in acidic water and acidic ethanolic extracts, respectively; 118 features were common to both extracts. These 118 features were compared to the 586 features that appeared in either of the basic extracts in order to search for the features detected specifically in acidic extracts or with higher intensity than in basic extracts. A total of 59 features were found to fit these criteria (Scheme 1).

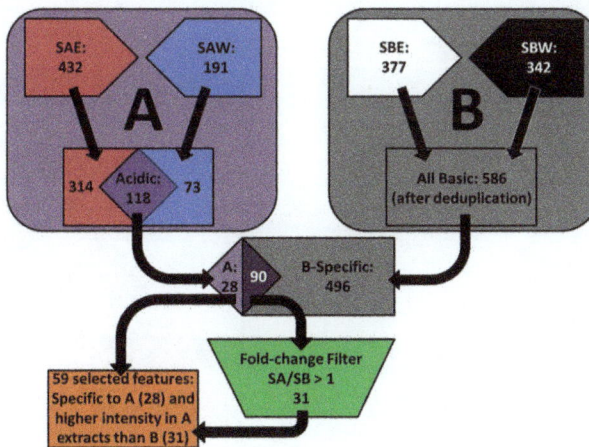

Scheme 1. Diagrammatic representation of the selection of features obtained from comparative analysis of chromatographic profiles between acidic and basic extracts in positive ionization mode ((**A**) Acidic extracts; (**B**) Basic extracts).

A similar procedure was then applied to the features obtained in negative ionization mode, resulting in five features of interest. The neutral mass of the 64 features obtained in both modes was queried against the METLIN Personal Metabolite Database (https://metlin.scripps.edu) and the customized NPACT database (http://crdd.osdd.net/raghava/npact) to match to compounds using a mass tolerance window of ≤10 ppm. The METLIN database contains 64,092 structures of endogenous and exogenous metabolites (as of June 2015) whereas the customized NPACT database consists of 1,574 entries of plant-derived natural compounds that exhibit anti-cancerous activity [18]. The METLIN-based search resulted in several candidates (a total of 880 hits for 64 features) and directed further investigation with MS fragmentation spectra to confirm the identities of bioactive compounds in the extracts (Table 2).

Table 2. Tentative identification of compounds which appear in acidic extracts exclusively or at greater extent than in basic extracts.

Formula	Experimental Mass	Error (ppm) *	Number of Candidates from Metlin Database	Putative Compounds from NPACT Database
$C_4H_5NO_3$	115.0269	0	1	-
$C_5H_7NO_3$	129.0428	1	7	-
$C_{14}H_{25}NO_3$	255.1835	0	1	-
$C_{14}H_{25}NO_3$	255.1846	4	1	-
$C_{15}H_{10}O_6$	286.0486	3	28	Fisetin, Kaempferol, Luteolin, Scullarein, Tetrahydroxyflavone
$C_{18}H_{33}NO_2$	295.252	2	1	-
$C_{15}H_{10}O_7$	302.0436	3	22	Quercetin; Morin;Viscidulin I
$C_{18}H_{19}NO_3$	307.2158	3	4	-
$C_{22}H_{43}NO$	337.335	1	3	-
$C_{22}H_{41}NO_2$	351.3145	2	1	-
$C_{21}H_{30}O_{10}$	442.1843	0	1	-
$C_{21}H_{20}O_{11}$	448.1009	0	74	Kaempferol β-D-glucopyranoside, Luteolin β-D-glucopyranoside
$C_{21}H_{20}O_{12}$	464.0958	0	56	Myricetin 3-O-rhamnoside
$C_{20}H_{26}O_{13}$	474.1373	0	1	-
$C_{28}H_{46}N_2O_4$	474.346	0	1	-
$C_{28}H_{46}N_2O_4$	474.3467	1	1	-
$C_{28}H_{46}N_2O_4$	474.3468	2	1	-
$C_{28}H_{48}N_2O_4$	476.3624	2	1	-
$C_{30}H_{48}N_2O_4$	500.3617	0	1	-
$C_{26}H_{55}O_8P$	526.3631	0	1	-
$C_{33}H_{50}O_5$	526.3631	5	1	-
$C_{34}H_{56}N_2O_4$	556.4245	0	1	-
$C_{29}H_{44}O_{12}$	584.2809	4	2	-
$C_{34}H_{48}O_8$	584.3322	4	3	-
$C_{27}H_{30}O_{15}$	594.158	0	120	-
$C_{27}H_{30}O_{16}$	610.153	0	109	Rutin
$C_{27}H_{30}O_{16}$	610.1531	0	109	-
$C_{28}H_{32}O_{16}$	624.1686	0	78	-
$C_{28}H_{32}O_{16}$	624.1689	0	78	-
$C_{34}H_{65}O_{13}P$	712.4155	1	1	-
$C_{33}H_{40}O_{19}$	740.2154	1	33	-
$C_{33}H_{40}O_{20}$	756.2107	0	81	-
$C_{33}H_{40}O_{21}$	772.2069	0	56	-
$C_{46}H_{77}O_9P$	804.5319	1	1	-

* Error (ppm): the difference between experimental mass and theoretical mass of compound.

Among the tentatively identified metabolites from the METLIN database, the most remarkable were the flavonoids or flavonoid glycosides (kaempferol, quercetin, rutin, manghaslin, nicotiflorin, clitorin, quercetin 7-galactoside, myricetin 7-rhamnoside, luteolin 3,7-diglucoside). These correlated with the results of the total flavonoid content in acidic extracts compared to basic extracts. Furthermore, the search within NPACT database revealed several compounds with previously reported anti-cancer activities including fisetin, kaempferol, luteolin, scullarein, tetrahydroxyflavone, quercetin, morin, viscidulin I, kaempferol β-D-glucopyranoside, luteolin β-D-glucopyranoside, myricetin 3-O-rhamnoside, and rutin [18]. Kaempferol and quercetin have been detected and quantified in *Carica papaya* leaves by gas chromatography-mass spectrometry analysis with quantities of 0.03 ± 0.001 mg/g and 0.04 ± 0.001 mg/g dry leaf, respectively [19]. Due to the presence of aromatic phenol groups, these flavonoids are considered weak acids; therefore they would be more readily extracted into an acidic environment in preference to a basic environment. Quercetin has been found to selectively affect the viability of SCC25 cells without causing toxicity to human gingival fibroblasts [20]. The mechanism by which quercetin inhibited the proliferation of SCC25 cells included both G1 phase cell cycle arrest and mitochondria-mediated apoptosis. Quercetin further decreased the migration and invasion of SCC25 cells in a dose-dependent manner [20]. For other flavonoids, many mechanisms of action on tumours have been identified such as inhibition of proliferation, apoptosis induction, carcinogen inactivation, impairment of invasion and angiogenesis [21,22]. The molecular mechanisms by which flavonoids exert these effects have been proposed to include the signalling pathways of PI3-kinase (phosphoinositide 3-kinase), Akt/PKB (protein-kinase B), tyrosine kinase P1KC (protein-1 kinase C) and MAP (mitogen-activated protein) kinase as reviewed elsewhere [23]. Using primary SCC cells and normal oral mucosa cells as a control, the flavonoid morin was shown to cause G2/M arrest without causing apoptosis, and to impact kinases AKT, JNK and p38 signaling pathways [24]. However, of the three kinases, AKT was the only one that was selectively inhibited in cancer cells compared to non-cancer cells, and the authors suggested that AKT might mediate the enhanced tumour sensitivity to morin [24].

Although our study presents limitations inherent to this type of research, in which database searching and matching based on accurate mass data alone provides numerous compound identities for each mass, the results revealed that papaya leaf acidic extracts contain numerous bioactive compounds with selective activities on SCC cells. Further studies are required to confirm the identities of these compounds by wider variety of isolation, purification and identification techniques (MS^n, NMR) and to investigate the possible mechanism of anticancer activities of *Carica papaya* leaf extracts.

3. Experimental Section

3.1. Chemicals and Reagents

Dulbecco's Modified Eagle's Medium (DMEM), DMEM-F12, penicillin/ streptomycin, trypsin, foetal bovine serum (FBS) were purchased from Invitrogen (Life Technologies, Mulgrave, VIC, Australia). 3-(4,5-Dimethylthiazol-2-yl)-2,5-diphenyltetrazolium bromide (MTT), dimethyl sulfoxide (DMSO), LC-MS grade ammonium bicarbonate and formic acid, Folin-Ciocalteu's phenol reagent, gallic acid (>97.5% purity), quercetin (98% purity) and epigallocatechin-3-gallate (EGCG) (99%) were obtained from Sigma-Aldrich (Castle Hill, NSW, Australia). HPLC grade methanol and acetonitrile were obtained from Merck (Darmstadt, Germany). Other chemicals such as hydrochloric acid, sodium hydroxide were of analytical grade purchased from Ajax Finechem (Cheltenham, VIC, Australia). Purified water was generated using a Milli-Q system (Millipore, Billerica, MA, USA).

3.2. Preparation of Papaya Leaf Extracts

Fresh *Carica papaya* leaves were collected from Tropical Fruit World (TFW), a privately-owned plantation orchard farm and research park in northern New South Wales, Australia (http://www.tropicalfruitworld.com.au/). Permission for the use of the *Carica papaya* leaves was granted by Aymon Gow, manager of TFW. The papaya plants in this facility are neither protected nor endangered species and had not been sprayed with any chemicals. The leaves were thoroughly washed under running tap water to remove any particulate matter, and then rinsed with deionized water to obtain clean leaves. A Christ Alpha 2-4 LD freeze-dryer was used to lyophilize the leaves at −60 °C and 0.1 mbar, for 24 h. Dried leaves were then ground into powder using a food processor (Oskar Mini, Sunbeam, NSW, Australia). The dried powder was portioned and stored at −80 °C until extraction.

A mass of 20 g of the freeze-dried leaf powder was extracted sequentially with ethanol and water in acidic or basic conditions in the order: basic ethanol, acidic ethanol, acidic water, basic water

(2 × 200 mL for each solvent; pH was maintained at pH 1–2 or 10–11 during the extraction process by adjusting with either 10% hydrochloric acid or 5 M sodium hydroxide solution). The obtained ethanolic extracts were concentrated at 30 °C and 120 rpm under vacuum by a rotary evaporator (IKA®RV 10, IKA-Werke, Staufen im Breisgau, Germany). The resulting concentrates of ethanolic extracts and water extracts were lyophilised using the freeze-dryer. All lyophilised fractions (Basic ethanol: SBE, Acidic ethanol: SAE, Acidic water: SAW, Basic water: SBW) were stored at −80 °C prior to analysis.

3.3. Cell Culture Conditions

SCC25 cells (ATCC® CRL-1628™, Manassas, VA, USA) were maintained in DMEM/F12 medium supplemented with 10% v/v heat-inactivated foetal bovine serum, 1% penicillin-streptomycin and 0.4 µg/mL hydrocortisone. HaCaT cells (a generous gift from Professor Fusenig) [25] were propagated in DMEM medium supplemented with 10% foetal bovine serum and 1% penicillin-streptomycin. The cells were grown in a humidified incubator at 37 °C in a 5% CO_2 atmosphere. Cells were passaged every 3 days and cultures were allowed to reach 70%–90% confluence before experiments were performed.

3.4. Cell Viability Assays

The effect of extracts on SCC25 and HaCaT viability was evaluated using the colorimetric tetrazolium dye procedure commonly referred to as the 3-(4,5-dimethylthiazol-2-yl)-2,5-diphenyltetrazolium bromide (MTT) assay developed by Mosmann with minor modification [26]. SCC25 or HaCaT cells were plated into 96-well plates at densities of 6×10^3 cells per well in 100 µL of DMEM/F12 (10% serum) and 3×10^3 cells well in 100 µL of DMEM (10% serum), respectively. Cells were incubated at 37 °C for 24 h and were subsequently treated for 48 h with 0.5% serum medium containing increasing concentrations (5–100 µg dry mass/mL) of extracts in ethanol (0.3% final concentration in medium). Control cells were exposed to an equivalent volume of ethanol (0.3% final concentration). Cells were then incubated in 100 µL of MTT-containing medium (0.2 mg/mL MTT in 0.5% serum medium) at 37 °C for an additional two hours. The medium was then removed and the formazan crystals trapped in cells were dissolved in 100 µL of DMSO by gentle shaking for 20 min on an orbital shaker. Absorbance of the solubilized product was measured at 595 nm using an Imark plate reader (BioRad, Hercules, CA, USA). The absorbance of cells exposed to medium containing 0.3% ethanol only was taken as 100% cell viability (*i.e.*, the control). The results are expressed as percent of the viability of control cells ± standard error of the mean (SEM) from 4–8 parallel determinations in three independent experiments ($n = 3$). Dose-effect analysis on SCC25 cells was performed by one-way analysis of variance (ANOVA) with Kruskal-Wallis test (as the data were non-parametric). Differences between the SCC25 and HaCaT cell lines, and interaction between cell line and extract effects were analysed by two-way ANOVA with Bonferroni post-tests. All statistical analyses were carried out using GraphPad Prism 6.0 (GraphPad Software Inc., San Diego, CA, USA, 2014).

3.5. Determination of Total Phenolic Content

Total phenolic content of ethanol and water extracts was determined using the Folin-Ciocalteu assay as described previously [27] with minor modifications. Briefly, 0.5 mL of diluted extract was mixed with 2.5 mL of freshly prepared Folin-Ciocalteu's phenol reagent, followed by the addition of 2 mL of 7.5% Na_2CO_3. The mixture was vortex mixed for 2 min and left in the dark at room temperature for 30 min. The absorbance was then measured at the maximum wavelength of 758 nm against a blank comprising 0.5 mL diluted extract, 2.5 mL water and 2 mL of 7.5% Na_2CO_3. Gallic acid was used as the standard, and results were expressed as gallic acid equivalents (GAE) in mg/g of dry weight of each extract.

3.6. Determination of Total Flavonoid Content

Total flavonoid content of the extracts was determined using a colorimetric assay developed previously [27]. Diluted extract was mixed with 2% AlCl₃ in ethanol in equal volume and absorbance was measured after 15 min at 425 nm, against the blank sample consisting of equal volume of dilute extract and ethanol without AlCl₃. Quercetin was used as the standard, and results were expressed as quercetin equivalents (QE) in mg/g of the dry weight of each extract.

3.7. UHPLC-ToF-MS Analysis

Chromatographic analysis of compounds in papaya leaf extracts was performed on an Agilent 1290 UHPLC system (Agilent Technologies, Santa Clara, CA, USA). Chromatographic separation was achieved on a 2.1 × 150 mm, 3.5 μm ECLIPSE PLUS C18 analytical column (Agilent) with guard protection. Mobile phase A was purified water containing 5 mM ammonium bicarbonate and 0.1% formic acid, adjusted to pH 7.0 ± 0.1 and mobile phase B was 95% acetonitrile and 5% water (*v/v*) containing 5 mM ammonium bicarbonate and 0.1% formic acid. The following gradient elution was adopted: 10% to 80% B for the first 42 min; 80% to 90% B from 42 to 45 min; 90% to 100% B from 46 to 48 min; held at 100% B from 48 to 50 min; returned to 10% B over the next 2 min, and the column re-equilibrated with 10% B for 10 min prior to the next injection. Thus the total chromatographic run time was 60 min. A flow rate of 0.2 mL/min was applied and 20 μL of sample was injected.

Each extract was run in triplicate in both positive and negative ionization mode. Mass spectrometric detection was performed on an Agilent 6520 high-resolution accurate mass quadrupole time-of-flight (Q-ToF) mass spectrometer equipped with a multimode source in both Electrospray Ionisation (ESI) and Atmospheric Pressure Chemical Ionization (APCI). Mass spectra were controlled using MassHunter acquisition software (Version B.02.01 SP3, Agilent Technologies, Santa Clara, CA, USA, 2010). The mass spectrometer was operated in the range of *m/z* 100–1700, at a scan rate of 0.8 cycles/second under the following conditions: capillary voltage 2500 V, nebulizer pressure 30 psi, drying gas flow 5.0 L/min, gas temperature 300 °C, fragmenting voltage 175 V, skimmer voltage 65 V. To ensure the desired mass accuracy of recorded ions, continuous internal calibration was performed during analysis with the use of reference ions—*m/z* 121.050873 (protonated purine) and *m/z* 922.009798 (protonated hexakis) in positive mode; in negative mode, ions with *m/z* 119.0362 (deprotonated purine) and *m/z* 966.000725 (formate adduct of hexakis) were used to correct for scan to scan variations.

3.8. MS Data Analysis

Data analysis was performed using Agilent MassHunter Qualitative software (Version B.05.00, Agilent Technologies, Santa Clara, CA, USA, 2012) with Molecular Feature Extractor (MFE) algorithms in concert with Mass Profiler Professional software (Version 12.1, Agilent Technologies, Santa Clara, CA, USA, 2012) to align features from the chromatograms of all samples from four extracts of papaya leaves. The following cut-off settings were employed: minimum peak filters of 500 counts, peak spacing tolerance of 0.0025 *m/z* plus 7.0 ppm, assigned charge states limited to a maximum of two, minimum compound filters of 3000 counts. The Molecular Feature Generator algorithm was utilised to generate putative molecular formulae from the following common elements C, H, N, O, P and S. Compound identification was carried out by using a Personal Compound Database Library (PCDL) (Agilent, Santa Clara, CA, USA) with the METLIN Personal Metabolite Database and a customized PCDL database which was compiled using the PCDL platform with compounds obtained from Naturally occurring Plant-based Anticancerous Compound-Activity-Target Database (NPACT) [18].

Acknowledgments: We would like to thank Aymon Gow and staff at Tropical Fruit World, New South Wales, Australia for their generosity in providing *Carica papaya* leaves for this study. Thao T. Nguyen is supported by Endeavour Postgraduate Award from Australian Government.

Author Contributions: Amitha K. Hewavitharana, Marie-Odile Parat, Mark P. Hodson and Paul N. Shaw conceived and designed the experiments; Thao T. Nguyen and Jenny Pan performed the experiments; Thao T. Nguyen, Amitha K. Hewavitharana, Marie-Odile Parat, Mark P. Hodson analyzed the data; Amitha K. Hewavitharana, Marie-Odile Parat, Mark P. Hodson and Paul N. Shaw contributed reagents/materials/analysis tools; Thao T. Nguyen wrote the paper; Amitha K. Hewavitharana, Marie-Odile Parat, Mark P. Hodson and Paul N. Shaw critically reviewed the manuscript. All authors read and approved the final manuscript.

Conflicts of Interest: The authors declare no conflict of interest.

References

1. Lucas, T.P. *The Most Wonderful Tree in the World, the Papaw Tree (Carica Papaia)*; Carter-Watson: Brisbane, QLD, Australia, 1914.

2. Tietze, H.W. *Papaya the Medicine Tree*, 3rd ed.; Harald Tietze Publishing: Bermagui, NSW, Australia, 2003.

3. Gallo, J.N. Papaya Leaf Tea for Cancer—Promoting, Researching, and Growing Papaya Leaf Tea. Available online: http://papayaleavesforcancer.com/ (accessed on 1 July 2015).

4. Dillan, J. Could Papaya Leaf be a Cancer Treatment? Available online: http://superfoodprofiles.com/papaya-leaves-cancer-treatment (accessed on 1 July 2015).

5. Clarks, D.H. Papaya leaf: The Anti-Cancer Treatment. Available online: http://www.huldaclarkzappers.com/php2/papayaleaf.php (accessed on 1 July 2015).

6. Barrett, M. Papaya Leaf Extract Is a Powerful Cancer Fighter. Available online: http://naturalsociety.com/papaya-leaf-extract-is-a-powerful-cancer-fighter/ (accessed on 1 July 2015).

7. Nguyen, T.T.T.; Shaw, P.N.; Parat, M.O.; Hewavitharana, A.K. Anticancer activity of *Carica papaya*: A review. *Mol. Nutr. Food Res.* **2013**, *57*, 153–164. [CrossRef] [PubMed]

8. Clayman, G.L.; Lee, J.J.; Holsinger, F.C.; Zhou, X.; Duvic, M.; El-Naggar, A.K.; Prieto, V.G.; Altamirano, E.; Tucker, S.L.; Strom, S.S. Mortality risk from squamous cell skin cancer. *J. Clin. Oncol.* **2005**, *23*, 759–765. [CrossRef] [PubMed]

9. Hewitt, H.; Whittle, S.; Lopez, S.; Bailey, E.; Weaver, S. Topical use of papaya in chronic skin ulcer therapy in jamaica. *West Indian Med. J.* **2000**, *49*, 32–33. [PubMed]

10. Mikhal'chik, E.V.; Ivanova, A.V.; Anurov, M.V.; Titkova, S.M.; Pen'kov, L.Y.; Kharaeva, Z.F.; Korkina, L.G. Wound-healing effect of papaya-based preparation in experimental thermal trauma. *Bull. Exp. Biol. Med.* **2004**, *137*, 560–562. [CrossRef] [PubMed]

11. Nayak, B.S.; Pereira, L.P.; Maharaj, D. Wound healing activity of *Carica papaya* L. In experimentally induced diabetic rats. *Indian J. Exp. Biol.* **2007**, *45*, 739–743. [PubMed]

12. Morimoto, C.; Dang, N.H. Compositions for cancer prevention, treatment, or amelioration comprising papaya extract. US Patent 20080069907, 20 March 2008.

13. Otsuki, N.; Dang, N.H.; Kumagai, E.; Kondo, A.; Iwata, S.; Morimoto, C. Aqueous extract of *Carica papaya* leaves exhibits anti-tumor activity and immunomodulatory effects. *J. Ethnopharmacol.* **2010**, *127*, 760–767. [CrossRef]

14. Rumiyati, S.A. Effect of the protein fraction of *Carica papaya* L. Leaves on the expressions of P53 and BCL-2 in breast cancer cells line. *Maj. Farm. Indones.* **2006**, *17*, 170–176.

15. Kaneshiro, T.; Suzui, M.; Takamatsu, R.; Murakami, A.; Ohigashi, H.; Fujino, T.; Yoshimi, N. Growth inhibitory activities of crude extracts obtained from herbal plants in the ryukyu islands on several human colon carcinoma cell lines. *Asian Pac. J. Cancer Prev.* **2005**, *6*, 353–358. [PubMed]

16. Manosroi, J.; Dhumtanom, P.; Manosroi, A. Anti-proliferative activity of essential oil extracted from thai medicinal plants on KB and P388 cell lines. *Cancer Lett.* **2006**, *235*, 114–120. [CrossRef] [PubMed]

17. Wang, G.H.; Chou, T.H.; Lin, R.J.; Sheu, J.H.; Wang, S.H.; Liang, C.H. Cytotoxic effect of the genus sinularia extracts on human SCC25 and HaCaT cells. *J. Toxicol.* **2008**, *2009*. [CrossRef]

18. Mangal, M.; Sagar, P.; Singh, H.; Raghava, G.P.; Agarwal, S.M. Npact: Naturally occurring plant-based anti-cancer compound-activity-target database. *Nucl. Acids Res.* **2013**, *41*, D1124–D1129. [CrossRef] [PubMed]

19. Canini, A.; Alesiani, D.; D'Arcangelo, G.; Tagliatesta, P. Gas chromatography-mass spectrometry analysis of phenolic compounds from *Carica papaya* L. Leaf. *J. Food Compos. Anal.* **2007**, *20*, 584–590. [CrossRef]

20. Chen, S.F.; Nien, S.; Wu, C.H.; Liu, C.L.; Chang, Y.C.; Lin, Y.S. Reappraisal of the anticancer efficacy of quercetin in oral cancer cells. *J. Chin. Med. Assoc.* **2013**, *76*, 146–152. [CrossRef] [PubMed]

21. Kanadaswami, C.; Lee, L.T.; Lee, P.P.H.; Hwang, J.J.; Ke, F.C.; Huang, Y.T.; Lee, M.T. The antitumor activities of flavonoids. *In Vivo* **2005**, *19*, 895–909.

22. Chahar, M.K.; Sharma, N.; Dobhal, M.P.; Joshi, Y.C. Flavonoids: A versatile source of anticancer drugs. *Pharmacogn. Rev.* **2011**, *5*, 1–12. [PubMed]

23. Hertzogi, D.; Tica, O. Molecular mechanism underlying the anticancerous action of flavonoids. *Curr. Health Sci. J.* **2012**, *38*, 145–149.

24. Brown, J.; O'Prey, J.; Harrison, P. Enhanced sensitivity of human oral tumours to the flavonol, morin, during cancer progression: Involvement of the Akt and stress kinase pathways. *Carcinogenesis* **2003**, *24*, 171–177. [CrossRef] [PubMed]

25. Boukamp, P.; Petrussevska, R.T.; Breitkreutz, D.; Hornung, J.; Markham, A.; Fusenig, N.E. Normal keratinization in a spontaneously immortalized aneuploid human keratinocyte cell line. *J. Cell Biol.* **1988**, *106*, 761–771. [CrossRef] [PubMed]

26. Mosmann, T. Rapid colorimetric assay for cellular growth and survival—Application to proliferation and cytotoxicity assays. *J. Immunol. Methods* **1983**, *65*, 55–63. [CrossRef]

27. Maisarah, A.; Amira, B.; Asmah, R.; Fauziah, O. Antioxidant analysis of different parts of *Carica papaya*. *Int. Food Res. J.* **2013**, *20*, 1043–1048.

Article

Withania somnifera Induces Cytotoxic and Cytostatic Effects on Human T Leukemia Cells

Eleonora Turrini [1], Cinzia Calcabrini [1], Piero Sestili [2], Elena Catanzaro [1], Elena de Gianni [1], Anna Rita Diaz [2], Patrizia Hrelia [3], Massimo Tacchini [4], Alessandra Guerrini [4], Barbara Canonico [2], Stefano Papa [2], Giovanni Valdrè [5] and Carmela Fimognari [1,*]

[1] Department for Life Quality Studies, Alma Mater Studiorum-University of Bologna, Corso d'Augusto 237, 47921 Rimini, Italy; eleonora.turrini@unibo.it (E.T.); cinzia.calcabrini@unibo.it (C.C.); elena.catanzaro2@unibo.it (E.C.); elena.degianni2@unibo.it (E.d.G.)

[2] Department of Biomolecular Sciences, University of Urbino Carlo Bo, Via Saffi 2, 61029 Urbino, Italy; piero.sestili@uniurb.it (P.S.); anna.diaz@uniurb.it (A.R.D.); barbara.canonico@uniurb.it (B.C.); stefano.papa@uniurb.it (S.P.)

[3] Department of Pharmacy and Biotechnology, Alma Mater Studiorum-University of Bologna, Via Irnerio 48, 40126 Bologna, Italy; patrizia.hrelia@unibo.it

[4] Department of Life Sciences and Biotechnology (SVeB)-LT Terra & Acqua Tech RU, University of Ferrara, Corso Ercole I d'Este 32, I-44121 Ferrara, Italy; massimo.tacchini@unife.it (M.T.); grrlsn@unife.it (A.G.)

[5] Department of Biological, Geological and Environmental Sciences (BiGeA), Alma Mater Studiorum-University of Bologna, Piazza di Porta S. Donato 1, 40126 Bologna, Italy; giovanni.valdre@unibo.it

* Correspondence: carmela.fimognari@unibo.it; Tel.: +39-0541-434-658

Academic Editor: Tzi Bun NG
Received: 30 December 2015; Accepted: 9 May 2016; Published: 12 May 2016

Abstract: Cancer chemotherapy is characterized by an elevated intrinsic toxicity and the development of drug resistance. Thus, there is a compelling need for new intervention strategies with an improved therapeutic profile. Immunogenic cell death (ICD) represents an innovative anticancer strategy where dying cancer cells release damage-associated molecular patterns promoting tumor-specific immune responses. The roots of *Withania somnifera* (*W. somnifera*) are used in the Indian traditional medicine for their anti-inflammatory, immunomodulating, neuroprotective, and anticancer activities. The present study is designed to explore the antileukemic activity of the dimethyl sulfoxide extract obtained from the roots of *W. somnifera* (WE). We studied its cytostatic and cytotoxic activity, its ability to induce ICD, and its genotoxic potential on a human T-lymphoblastoid cell line by using different flow cytometric assays. Our results show that WE has a significant cytotoxic and cytostatic potential, and induces ICD. Its proapoptotic mechanism involves intracellular Ca^{2+} accumulation and the generation of reactive oxygen species. In our experimental conditions, the extract possesses a genotoxic potential. Since the use of *Withania* is suggested in different contexts including anti-infertility and osteoarthritis care, its genotoxicity should be carefully considered for an accurate assessment of its risk–benefit profile.

Keywords: *Withania somnifera*; apoptosis; cell cycle; leukemia; oxidative stress; immunogenic cell death; genotoxicity

1. Introduction

Cancer causes millions of deaths every year. Only in 2012, cancer deaths have reached 8.2 million. In 2030, 12.6 million cancer deaths have been estimated [1].

Cancer originates from multiple alterations induced by a direct interaction between toxic agents and DNA. This triggers gene and chromosome mutations. Altered expressions of oncogenes and tumor suppressor genes are found in different cancer types. The consequence is an uncontrollable proliferation mediated by growth signals released from the tumor cells themselves, resistance against antigrowth signals, and inhibition of apoptosis. Furthermore, cancer expansion is helped by the release of various growth factors which lead to the formation of new blood vessels that provide nutrients and oxygen thus favoring cancer spread and metastasis dissemination [2]. Despite the progress made in anticancer research, traditional cytotoxic chemotherapy continues to serve as the basis for the current standard therapeutic regimen. This is characterized by an elevated intrinsic toxicity, mainly due to its poor selectivity for cancer cells. Furthermore, cancer cells' ability to develop drug resistance represents a major problem in anticancer therapy [3]. Thus, there is a compelling need for new intervention strategies with an improved therapeutic profile. Cancer cells create a favorable microenvironment allowing them to survive, proliferate, and counteract immunosurveillance. A promising anticancer strategy could be represented by the use of cytotoxic drugs that are not only able to induce tumor cell death, but also promote tumor-specific immune responses, potentially preventing tumor progression and relapse [4]. Very recent studies have introduced the concept of immunogenic cell death (ICD), a modality of cell death where, after the exposure to some cytotoxic agents, dying cancer cells release endogenous damage-associated molecular patterns (DAMPs) including calreticulin, heat shock protein (Hsp)-70 and Hsp-90 recognized by antigen-presenting cells such as dendritic cells (DCs). This is followed by T-cell-mediated adaptive immunity [5].

Natural products represent a rich source of biologically active compounds that can be able to interact simultaneously with different targets involved in cell growth, cell differentiation, and apoptosis regulation [6]. In 2015, the Food and Drug Administration (FDA) released an updated guidance on botanical drug development. Unlike drugs that are constituted by a single active ingredient, botanical drugs have a heterogeneous nature, which may lead to uncertainty in relation to their active constituents. The number of botanical products submitted to the FDA is particularly high in the oncological area [7]. The first botanical drug in this area is Polyphenon E, a standardized extract obtained from the leaves of green tea (*Camellia sinensis*), approved by the FDA in 2007 for treatment of genital warts linked to human papilloma viruses. The well-defined make-up, standardization, and cheap cost make Polyphenon E a very interesting candidate for human clinical studies. Polyphenon E is currently in several trials as a chemopreventive and chemotherapeutic agent against chronic lymphocytic leukemia, bladder and lung cancers (phase II), and in breast cancer (phase I) [8–10].

Many recent studies focus on the potential anticancer effect of crude extracts from plants used in traditional medicine and their isolated compounds. The roots of *Withania somnifera* (*W. somnifera*), a plant originating from Asia and South Africa [11], are used in the Indian traditional medicine [12]. A wide range of biological activities is reported for *W. somnifera* including anti-inflammatory [13], immunomodulating [14], neuroprotective [15], and anticancer activities [16]. The present study is designed to explore the antileukemic activity of the DMSO extract obtained from the roots of *W. somnifera* (WE). Particular emphasis is given to the role of reactive oxygen species (ROS) in its anticancer effect. With the aim to extend the potential clinical impact of *Withania*, we investigated its ability to induce ICD and assessed on a preliminary basis the risk–benefit profile associated with the use of this plant through the analysis of its genotoxic potential.

2. Results

2.1. WE Contains Withaferin A (WFA), Whitanolide A (WDA), Withanolide B in Trace Amount

We detected and quantified WFA and WDA (Table 1), which are among the most representative markers of *Withania somnifera* [17]. WFA was also described as highly soluble in DMSO, confirming our results. Withanolide B was instead under the Limit Of Quantification (LOQ = 4.36 ± 0.65 µg/mL) and withanone undetectable.

Table 1. Quantification of withaferin A (WFA) and withanolide A (WDA).

Compound	Amount (µg/mL)	LOD	LOQ	Amount (mg/g of Dried Extract)	Recovery %
WFA	113.65 ± 2.84	6.54 ± 0.11	19.81 ± 0.63	5.68 ± 0.14	96.85 ± 1.98
WDA	39.42 ± 1.44	1.64 ± 0.07	4.96 ± 0.26	1.97 ± 0.07	110.57 ± 2.11
withanolide B	tr	2.03 ± 0.34	6.36 ± 0.65	-	-
withanone	-	1.99 ± 0.29	15.95 ± 1.18	-	-

LOD: limit of detection, LOQ: limit of quantification, tr = trace.

2.2. WE Induces Apoptosis and Alters Cell-Cycle Residence

WE causes a dose-dependent reduction of cell viability. For example, after 24 h treatment of Jurkat cells with 1.6 mg/mL of WE, the percentage of viable cells was 64.4% and at 3.2 mg/mL cells viability achieved 16.6%. The calculated IC_{50} value (the inhibitory concentration causing cell toxicity by 50% following one cell-cycle exposure) was 2.3 mg/mL. Concentrations similar or smaller than the IC_{50} were used in the following experiments.

Further analyses were carried out to discriminate whether the inhibitory effect of WE on cell viability was the result of apoptotic cell death. After 6 h of treatment at 0.4 and 0.8 mg/mL, WE significantly increased the percent of apoptotic cells (3.4- and 4.1-fold increase, respectively, *versus* untreated cells). After 24 h of treatment, the percent of apoptotic cells was statistically significant starting from 0.4 mg/mL, where $33.1\% \pm 3.7\%$ of apoptotic events was observed *versus* $3.1\% \pm 0.2\%$ of untreated cells (Figure 1). An increase in necrotic events was also recorded starting from 0.80 mg/mL ($11.6\% \pm 1.9\%$ *versus* $1.7\% \pm 0.2\%$ of untreated cells). At the highest tested concentration of WE (1.6 mg/mL), both apoptotic and necrotic events markedly increased, but the percentage of apoptotic cells was significantly higher than that of necrotic cells (53.2% *versus* 28.2%, respectively) (Figure 1A). When cells were treated with WFA, WDA or their association, we observed an increase in the fraction of apoptotic cells only for WFA at all the concentrations tested (Figure 1B). The proapoptotic effect of the association WFA plus WDA was very similar to that of WFA (Figure 1B). In Figure 1C, we compared the fold increase in the percent of apoptotic cells recorded after treatment with WE, WFA or WFA plus WDA. The concentrations of WFA and WDA are those found in the extract at 0.20, 0.40 and 0.80 mg/mL. Even if WFA and WFA plus WDA possess a proapoptotic effect, the effect of WE is significantly higher than that observed for WFA or the association.

In the following experiments, we highlighted the cytostatic effect of WE. After 24 h treatment at increasing concentrations of WE, we observed an increasing number of cells in G2/M phase starting from 0.1 mg/mL ($39.6\% \pm 0.1\%$ *versus* $22.4\% \pm 1.2\%$ of untreated cells), accompanied by a decrease in cells in phase G0/G1 ($45.7\% \pm 0.1\%$ *versus* $63.0\% \pm 2.2\%$ of untreated cells) (Figure 2). WE showed the same trend up to 0.4 mg/mL, where we detected an increase in cells in G2/M phase ($30.3\% \pm 0.9\%$) and a decrease in cells in G0/G1 phase ($50.1\% \pm 2.2\%$). At the highest concentrations tested, the cell-cycle distribution was similar to that of untreated cells (Figure 2).

2.3. WE Increases Intracellular Ca^{2+} ($[Ca^{2+}]_i$)

We explored the ability of WE to modulate $[Ca^{2+}]_i$ on viable cells after 6 and 24 h of Jurkat treatment with WE. The extract increased $[Ca^{2+}]_i$ in a dose- and time-dependent manner. At 6 h, $[Ca^{2+}]_i$ was significantly enhanced only at the highest concentration (0.4 mg/mL) tested [837.5 ± 86.9 MFI (mean fluorescence intensity) *versus* 410.5 ± 12.4 MFI of untreated cells] (Figure 3). After 24 h of WE, we recorded a significant increase in $[Ca^{2+}]_i$ at all tested concentrations, starting from 0.1 mg/mL (535.5 ± 61.5 MFI *versus* 399.7 ± 26.9 MFI of untreated cells) and becoming 3.7 fold higher than control at the highest tested concentration (1530 ± 27.7 MFI). Dead cells were analyzed separately as unique cluster. We observed an increase in $[Ca^{2+}]_i$ (data not shown) that confirms the involvement of Ca^{2+} in the antileukemic effect of WE.

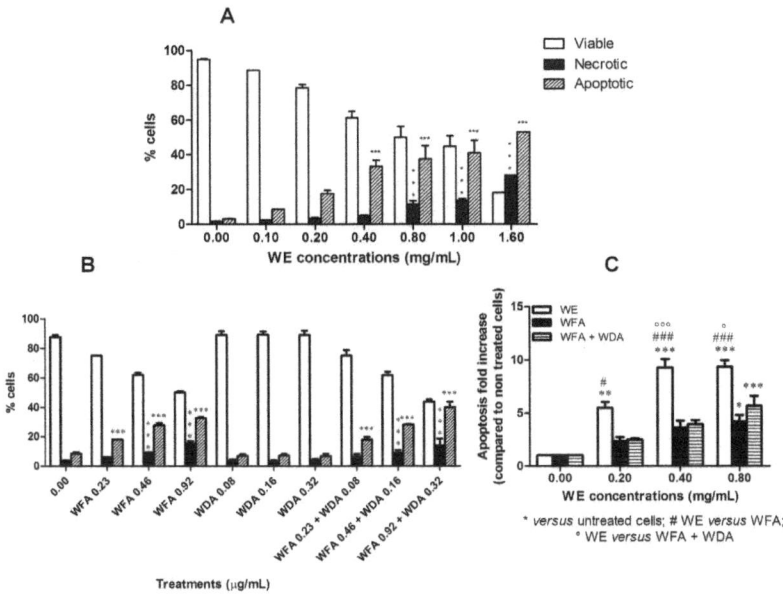

Figure 1. Percentage of viable, necrotic and apoptotic cells after 24 h treatment of Jurkat cells with increasing concentrations of: DMSO extract obtained from the roots of *W. somnifera* (WE) (**A**); and withaferin A (WFA), withanolide A (WDA) or WFA plus WDA (**B**). Fold increase in the percent of apoptotic cells after treatment with different concentrations of WE, WFA, or WFA plus WDA (**C**). * $p < 0.05$; ** $p < 0.01$; *** $p < 0.001$ *versus* untreated cells.

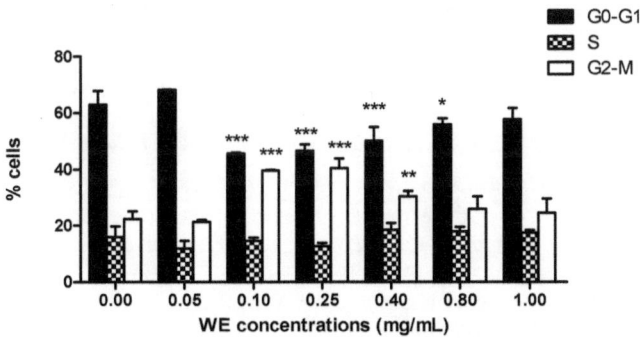

Figure 2. Cell-cycle distribution following 24 h treatment of Jurkat with increasing concentrations of WE. * $p < 0.05$; ** $p < 0.01$; *** $p < 0.001$ *versus* untreated cells.

2.4. WE Induces Oxidative Stress

WE extract increased ROS production in a dose-dependent manner in Jurkat cells (Figure 4A). Most of the ROS were generated between 3 and 6 h of incubation. After 6 h treatment with WE, 0.8 and 1.6 mg/mL of WE led to ROS levels similar or even higher than those promoted by a mildly toxic dose of H_2O_2 (0.1 mM for 15 min), included as a positive control. After longer times of treatment (18 and 24 h), ROS generation reached a plateau (data not shown). *N*-acetylcysteine (NAC) and o-phenantroline (o-Phe) significantly inhibited the ROS generation induced by WE (0.8 mg/mL for 6 h), while rotenone (Rot) was unable to afford a protective effect (Figure 4B).

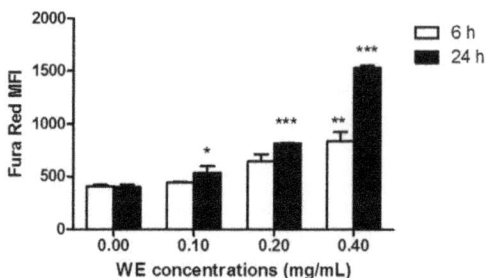

Figure 3. Fraction of living cells with increased $[Ca^{2+}]_i$ following 6 and 24 h exposure to increasing concentrations of WE. * $p < 0.05$; ** $p < 0.01$; *** $p < 0.001$ *versus* untreated cells.

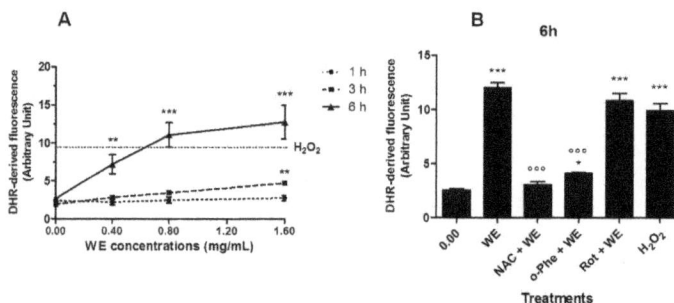

Figure 4. Reactive oxygen species (ROS) generation in WE-treated cells: (**A**) Jurkat exposed to increasing concentrations of WE for 1 h, 3 h or 6 h. Cells treated with H_2O_2 0.1 mM for 15 min represent the positive control (dotted line parallel to *x*-axis). (**B**) Cells were treated for 6 h with WE 0.8 mg/mL in the absence or presence of o-phenanthroline (o-Phe, 10 µM), rotenone (Rot, 2 µM) or *N*-acetylcysteine (NAC, 10 mM). Cells treated with H_2O_2 0.1 mM for 15 min represent the positive control. * $p < 0.05$, ** $p < 0.01$, *** $p < 0.001$ versus control, and $^{\circ\circ\circ}$ $p < 0.001$ versus WE.

2.5. Co-Treatment of Cells with WE and NAC Significantly Decreases WE-Induced Apoptosis

Because of the crucial role of ROS in the bioactivity of WE, we investigated whether the alteration of the redox state induced by NAC treatment could play a role in the apoptosis induced by WE. We observed a significant decrease in the WE-induced apoptotic events following 24 h of Jurkat co-treatment with WE plus NAC (10 mM). The WE-induced apoptotic events were significantly reduced from 35% after 0.40 mg/mL of WE to 10% after WE plus NAC and from 38% after 0.80 mg/mL of WE to 12% after WE plus NAC (Figure 5).

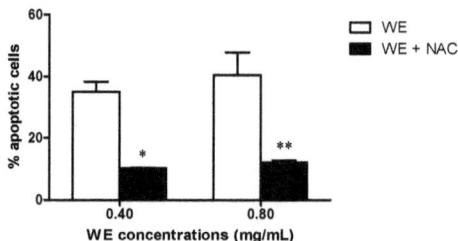

Figure 5. Apoptotic events after 24 h of Jurkat treatment with WE in the absence and presence of *N*-acetyl cysteine (NAC) (10 mM). * $p < 0.05$; ** $p < 0.01$ *versus* WE.

Moreover, we co-treated cells with WE and L-asparagine (1–2 mM): under this condition, we did not record any modulation of the proapoptotic potential of WE (data not shown).

2.6. WE Induces ICD

Based on the cytotoxic activity of WE and its ability to increase intracellular ROS and Ca^{2+} levels, we preliminarily explored the capacity of WE to induce ICD. To this aim, the exposure of some DAMPs on the extracellular membrane of Jurkat cells was examined. After 6 h of treatment, we did not observe any effect of WE on calreticulin translocation (data not shown). Longer treatment times (24 h) caused calreticulin translocation on the extracellular membrane (Figure 6A,D), with a mean fluorescence of 8.39 at 0.2 mg/mL and 8.72 at 0.4 mg/mL compared to 6.83 of the control (Figure 6A). Similarly, cells treated with WE showed an increase in both Hsp-70 and Hsp-90 expression only after 24 h of treatment (Figure 6E,F, respectively). As an example, at 0.2 mg/mL Hsp-70 fluorescence was 57.92 compared to 42.17 of the control and Hsp-90 mean fluorescence was 144.14 compared to 58.45 of untreated cells; at 0.4 mg/mL, the fluorescence of both Hsp-70 and Hsp-90 increased to 85.22 and 151.09, respectively (Figure 6B,C, respectively). Finally, we measured the release of adenosine triphosphate (ATP) from dying cells after 6 and 24 h of treatment with WE. At 6 h, we did not record any modulation of ATP levels (data not shown). After 24 h, we observed a significant increase in ATP levels at both the tested concentrations of the extract (2- and 2.45-fold increase, respectively) (Figure 6G).

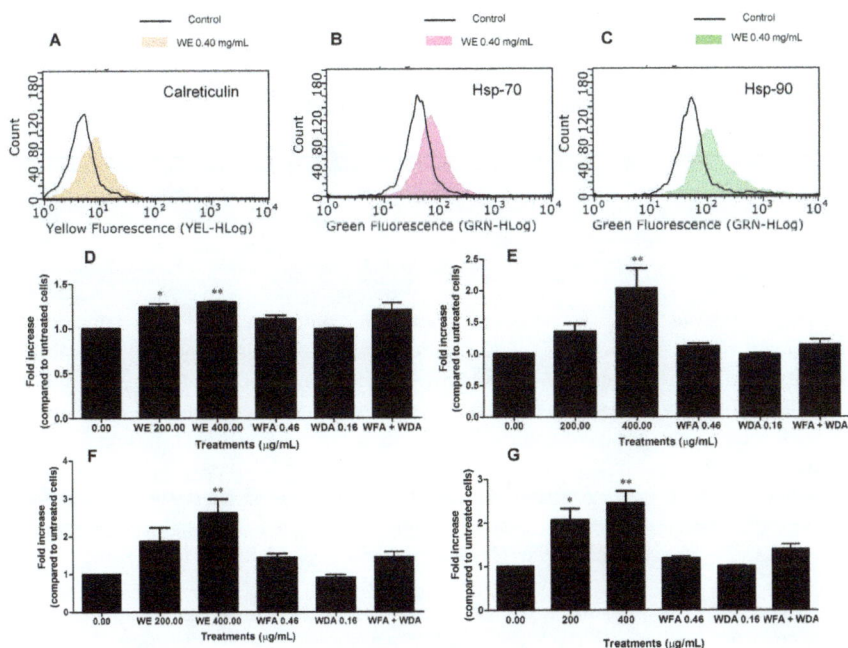

Figure 6. Fluorescence hystograms of: immunolabeled calreticulin (**A**); Hsp-70 (**B**); and Hsp-90 (**C**). Modulation of the expression of: calreticulin (**D**); Hsp-70 (**E**); Hsp-90 (**F**); and of ATP release (**G**) after treatment with WE, WFA, WDA or WFA plus WDA. Histograms are representatives of three independent experiments. * $p < 0.05$, ** $p < 0.01$ *versus* untreated cells.

We also tested the induction of ICD by the two main constituents of WE extract (*i.e.*, WFA and WDA), used at the concentrations found in the WE extract at 0.40 mg/mL. Treatment with WFA and

WDA alone or in association for 24 h did not cause a statistically significant modulation of calreticulin translocation, Hsp-70 and Hsp-90 expression or ATP release (Figure 6D–G).

2.7. WE Induces DNA Damage

To evaluate the ability of WE to induce DNA damage, H2A.X phosphorylation was analyzed. H2A.X phosphorylation at Ser 139 represents a sensitive marker for DNA strand breakage [18]. WE induced a dose-dependent increase in H2A.X phosphorylation, which was eight times higher than untreated cells at the highest tested concentration (0.80 mg/mL). This increase was similar to the phosphorylation induced by etoposide 10 μM, used as positive control (Figure 7).

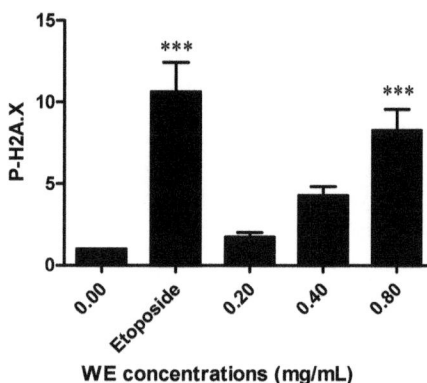

Figure 7. Relative expression of phosphorylated H2A.X (P-H2A.X) induced by WE in Jurkat cells after 6 h of treatment. Etoposide (10 μM) was used as positive control. *** $p < 0.001$ *versus* untreated cells.

3. Discussion

In this study, we demonstrated the *in vitro* antileukemic effect of the root extract of *W. somnifera* in a T-lymphoblastoid cell line. The high-performance liquid chromatography (HPLC) analysis performed on WE revealed the presence of WFA, WDA and to a lesser extent withanolide B. Withanolides, in particular WFA and its acetyl derivative, are highly bioactive and show anticancer activity [19–21]. Of note, its dihydroderivative is not active, thus suggesting that an unsaturated lactone moiety in ring A of WFA is important for its biological activity. In our experimental settings, WE significantly induced apoptosis in a remarkable proportion of cells. Moreover, it blocked cell proliferation through an accumulation of cells in the G2/M phase starting from the lowest tested concentrations. Cell-cycle dysregulation represents a hallmark of cancer [22] and targeting the checkpoint signaling pathway, which usually leads to an arrest at G1/S or G2/M boundaries, is an effective therapeutic strategy [23]. Our results confirm the antiproliferative and proapoptotic effect reported for withanolides and for a methanolic crude *Withania* leaf extract in different leukemia cell lines [24,25]. However, the IC$_{50}$ calculated for the above mentioned methanolic leaf extract was much lower than that calculated in our study. The difference could be imputable to the different part of plant (root *versus* leaf) used and/or the method of extraction (the methanolic leaf extract was subjected to a sequential solvent extraction, which progressively concentrated the active components of *Withania* leaves).

Different mechanisms can be involved in the proapoptotic activity of our WE. Numerous studies reported that intracellular Ca^{2+} mobilization plays a crucial role in apoptosis [26,27] and that calcium ionophores exhibit proapoptotic activity [28]. Since WE exhibited a marked proapoptotic ability, we measured [Ca^{2+}]$_i$ and demonstrated that further to the treatment of Jurkat cells with WE, [Ca^{2+}]$_i$ significantly increased. The underlying mechanisms need to be explored, however we can hypothesize that the proapoptotic mechanism of WE involves intracellular Ca^{2+} accumulation.

Fruits of *W. somnifera* contain different enzymes including L-asparaginase, which catalyzes the conversion of the aminoacid asparagine to aspartate and ammonia. Through this mechanism, L-asparaginase depletes the cellular levels of asparagine and induces the death of leukemic cells that are unable to synthesize asparagine. L-asparaginase exhibits cytotoxic effects on patient-derived leukemic blasts [29]. A specific phytochemical analysis should be performed to detect the presence of L-asparaginase in our extract. However, in our experimental conditions, the co-treatment of cells with WE plus L-asparagine did not affect the proapoptotic potential of WE. The latter finding may suggest that our extract does not contain L-asparaginase. Thus, it is conceivable that the presence of L-asparaginase does not play a critical role in the cytotoxic activity of WE.

As already mentioned, the main component of our WE is WFA. Many studies have demonstrated the role of WFA inducing oxidative stress as an anticancer strategy on different tumor cell lines, such as prostate cancer, breast cancer, pancreatic cancer, leukemia, and melanoma [30–34]. ROS-mediated apoptosis by WFA was shown to depend on both intrinsic and extrinsic pathways. Mitochondrial membrane potential loss, release of cytochrome c and translocation of Bax, as well as increase in caspase-8 activity were observed together with the decrease of Bid, as crosstalk between intrinsic and extrinsic pathways [32]. Accordingly, we demonstrated that WE increased intracellular ROS levels. The remarkable inhibition of the proapoptotic potential of WE after co-treatment with NAC confirms the key role of ROS production in the apoptosis induced by WE. Similar data were obtained in both estrogen receptor (ER) positive- and ER negative-breast cancer cell lines, where the apoptotic effect of WFA was blunted by the presence of antioxidants [30].

Notably, selectivity by WFA towards cancer cells was observed in pancreatic and breast cancer cells, as compared to normal human fibroblasts and normal human mammary epithelial cell line [30,34]. ROS levels in cancer cells are close to the threshold and a ROS-mediated apoptotic mechanism represents an established indicator of cancer selectivity for an anticancer compound [35].

WE extract caused a dose-dependent ROS generation in Jurkat cells. This finding is in agreement with previous studies reporting the ROS-generating ability of a similar WE extract [32] and some of its components such as WFA [32,33,36–38]. ROS generation induced by WE extract reached a plateau after 6 h of incubation. This suggests that the extract rapidly induces a pro-oxidative status in intoxicated cells. Similar results were obtained by Malik [32], who found a significant ROS increase following 1 to 3 h of exposure to WFA [38]. The co-incubation of cells with WE extract and NAC, an established ROS scavenger, quenched the ROS generation induced by WE. This finding is in conformity with previous data that reported that NAC attenuates *Withania*-induced ROS production in several cell lines [32,36–38]. Co-incubation with o-Phe, an iron chelator that breaks Fenton reaction and stops ROS generation [39], attenuated ROS generation to a similar extent as NAC. These data strengthen the notion that WE causes the cellular formation of ROS. Since analyses performed on melanoma cell lines treated with WFA recorded mitochondrial ROS generation [33], we investigated whether the mitochondrial respiratory chain is involved in this process. Rot, a prototypical Complex I inhibitor [40], did not affect WE-induced ROS production, suggesting that this complex is not involved in this process and that further studies will be needed to individuate the exact site of ROS production.

ROS production and endoplasmic reticulum (ER) stress are critical event promoting ICD, which is also associated with the expression and/or release of DAMPs [41]. For example, calreticulin is a DAMP usually located on the lumen of ER and translocated on the extracellular membrane in case of ER stress [41]. ER regulates many cellular events including $[Ca^{2+}]_i$ levels. Alterations in Ca^{2+} homeostasis causes ER stress [42]. In our experimental settings, WE treatment causes ROS production and increases $[Ca^{2+}]_i$ levels. Accordingly, to the best of our knowledge, we demonstrated for the first time the ability of *Withania* to induce ICD starting from the lowest tested concentrations, as indicated by the up-regulation of calreticulin, Hsp-70, Hsp-90, and ATP release. The expression of these molecules increases the immunogenic profile of tumor cells, thus promoting the innate immune system response [43]. However, WFA and WDA that represent the two main constituents of our WE extract did not induce ICD either alone or in association. Recent evidence shows that WFA does not alter the expression of Hsp-90 either in lymphoma or in pancreatic cells, and inhibits Hsp-90 with an ATP-independent mechanism [44–46]. The anticancer activity of WFA depends on the inhibition of critical kinases and cell-cycle regulators controlled by Hsp-90 [44,45]. Hsp-90 is a molecular chaperone involved in regulating protein folding and modulating a number of oncogenic client proteins playing a critical role in oncogenesis and cancer progression. Hsp-90 depends upon different co-chaperones for its function. WFA blocks the association of Hsp-90 to Cdc37, *i.e.*, its co-chaperone, thus acting as a potent Hsp-90-client modulating agent [46]. The different activity of the two withanolides and WE could be imputable to the complex nature of WE. In other words, the combined effects of the bioactive molecules of the extract could differently influence DAMPs' expression.

An immunostimulatory activity of an aqueous/alcoholic (1:1) root extract of *Withania* has been reported on BALB/c mice and on *ex vivo* and *in vitro* macrophages [47]. The immunostimolatory activity of *Withania* together with the induction of ICD represents a promising strategy for the generation of a tumor-specific response.

Finally, we analyzed the genotoxic potential of WE. Our results showed that the treatment of Jurkat cells with WE significantly boosts H2A.X phosphorylation, which is an index of the ability of a compound to interact with DNA thus triggering a genotoxic lesion. However, some recent *in vivo* studies reported the lack of genotoxicity of one of the most important constituents of *Withania, i.e.*, WFA, and demonstrated its ability to provide protection against the 7,12-dimethylbenz(a)anthracene-induced genotoxicity [48,49]. As with the different behavior of WE and withanolides in ICD induction, the different genotoxic profile of WE and WFA could be due to the matrix effects. Genotoxic studies on complex products of natural origin are usually performed on single phytochemicals rather than on the product in its complexity. The matrix effect can cause an incomplete release of a key constituent from the vegetal matrix or modulate its bioavailability. This means that the use of toxicity data concerning the pure phytochemical are unsuited for the purposes of assessing the risk derived from the use of the same phytochemical within the complex vegetal matrix [2]. The use of *Withania* is suggested in different contexts including naturopathic care for anxiety [50], anti-infertility care [51], and osteoarthritis care [52]. Its genotoxicity should be carefully considered for an accurate assessment of its risk–benefit profile. It is important to note that the H2A.X phosphorylation test used in our study is able to detect only premutational, thus reparable DNA lesions. For this reason, further experiments are needed to define the net and actual mutagenic effect of the lesions caused by WE and to directly relate the DNA damage to the mutagenic effect.

4. Experimental Section

4.1. WE Preparation

Withania somnifera roots were collected during the balsamic period (summer) and authenticated by Dr. Paolo Scartezzini, Maharishi Ayurveda Product Ltd., Noida, India. The quality control was performed by Vedic Herbs s.r.l. (Caldiero, VR, Italy), which gifted us with a sample of root powder (voucher #12/11). The extract was prepared by mixing 10 g of *Withania* root powder with 100 mL of DMSO. The extract was vortexed for 15 min at room temperature and centrifuged to discard any insoluble part. The experiments and the HPLC analysis were performed using this stock solution of 100 mg/mL.

4.2. HPLC Analysis

WE was subjected to RP-HPLC-DAD analysis to identify and quantify the main phytomarkers. The reference compounds WFA, WDA, withanolide B, and withanone were purchased from Extrasynthese, Lyon, France. WFA and WDA were used as external standards to set up and calculate appropriate calibration curves. The analyses were performed using a Jasco modular HPLC (model PU 2089, Jasco Corporation, Tokyo, Japan,) coupled to a diode array apparatus (MD 2010 Plus) linked to an injection valve with a 20 μL sampler loop. The column used was a Kinetex XB-C18 (5 μm,

15 cm × 0.46 cm) with a flow rate of 0.6 mL/min. The analyses were performed at 25 °C with mobile phase and gradient chosen according to literature [17].

Following chromatogram recording, sample peaks were identified by comparing their ultraviolet (UV) spectra and retention time with those of the pure standards. Dedicated Jasco software (PDA version 1.5, Jasco Corporation, 2004) was used to calculate peak area by integration.

4.3. Validation

The individual stock solutions of each phytomarkers were prepared in ethanol or acetonitrile. The calibration curves of the considered compounds were prepared within different range: 500–50 μg/mL for WFA, and 100–10 μg/mL for WDA. Each calibration solution was injected into HPLC in triplicate. The calibration graphs were provided by the regression analysis of the peak area of the analytes *versus* the related concentrations. The analysis of the extract was performed under the same experimental conditions. The obtained calibration graphs allowed the determination of the concentration of the phytomarkers inside the extract.

Limit Of Detection (LOD) and LOQ were calculated following the approach based on the standard deviation of the response and the slope for WFA and WDA, on signal and noise ratio for withanolide B and withanone, as presented in the "Note for guidance on validation of analytical procedures: text and methodology", European Medicine Agency ICH Topic Q2 (R1). The accuracy was reported as percent of recovery and was estimated by adding known amount of analyte in the studied sample.

4.4. Cell Cultures

Human T-lymphoblastoid cells (Jurkat) were provided from LGC standards (LGC Group, Middlesex, UK). Cells were grown in suspension in Roswell Park Memorial Institute (RPMI) 1640 supplemented with 10% heat-inactivated bovine serum, 1% penicillin/streptomycin solution, and 1% L-glutamine solution (all obtained from Biochrom, Merck Millipore, Darmstadt, Germany). Cells were incubated at 37 °C with 5% CO_2. To maintain exponential growth, the cultures were diluted to never exceed the maximum suggested density of 3×10^6 cells/mL.

4.5. Cell Treatment

Cells were treated with increasing concentrations of WE (0.0–1.6 mg/mL) for 1, 3, 6 or 24 h, according to the experimental requirements, or with WFA, WDA or WFA plus WDA for 24 h. WFA and WDA were tested at the concentrations found in the WE extract at 0.2, 0.4 and 0.8 mg/mL: 0.23–0.92 μg/mL for WFA; 0.08–0.32 μg/mL for WDA. Etoposide 10 μM and hydrogen peroxide 0.1 mM were used as positive controls. In some experiments, a co-treatment of WE with NAC or L-asparagine was performed.

4.6. Analysis of Cell Viability and Induction of Apoptosis

To determine cells' viability, Guava ViaCount Reagent (Merck Millipore, Darmstadt, Germany) was used according to manufacturer's instructions. Briefly, cells were appropriately diluted with the reagent containing 7-amino-actinomycin D (7-AAD) and incubated at room temperature in the dark for 5 min before detection with flow cytometer. Furthermore, to discriminate between necrotic and apoptotic events, Guava Nexin Reagent (Merck Millipore) was used. Through the use of 7-AAD and annexin V-phycoerythrin, the assay allows the discrimination of apoptotic and necrotic events. Cells were incubated with the reagent for 20 min at room temperature in the dark and then analyzed via flow cytometry. IC_{50} was calculated by interpolation from dose–response curve. Concentrations $\leqslant IC_{50}$ were used in the subsequent experiments.

4.7. Cell-Cycle Analysis

After treatment with WE for 24 h, cells were fixed with 70% ice-cold ethanol and, after washing, suspended in 200 μL of Guava Cell Cycle Reagent (Merck Millipore), containing propidium iodide. At the end of incubation at room temperature for 30 min in the dark, samples were analyzed via flow cytometry.

4.8. Measurement of $[Ca^{2+}]_i$

After WE treatment for 6 or 24 h, $[Ca^{2+}]_i$ was analyzed by using Fura Red™, AM (Thermo Fisher Scientific, Carlsbad, CA, USA), according to manufacturer's instructions. Briefly, after treatment, cells were incubated with the dye that freely permeates the cytoplasmic membrane but, once inside the cells, is hydrolyzed by the intracellular esterases and trapped into the cells. The fluorescence of this molecule is enhanced once it binds Ca^{2+}. To determine the optimal concentration of dye, a titration of Fura Red™, AM was performed by loading Jurkat cells with a range of concentrations recommended by the manufacturer (1–10 μM). The exposure of cells to Fura Red can cause cell death [53]. Thus, the use of the lowest concentration of Fura Red is recommended. Following this experimental phase, the concentration of 1 μM was adopted.

To detect intracellular calcium levels, Jurkat cells were incubated at 37 °C for 30 min in PBS without calcium and magnesium. This buffer condition allows to detect the intracellular calcium stores and exclude the secondary increase in $[Ca^{2+}]_i$ due to Ca^{2+} entry [54]. Moreover, the removal of external Ca^{2+} reduces the non-specific fluctuations in $[Ca^{2+}]_i$ normally observed during the first 20–30 s of sample acquisition via flow cytometry. Results are expressed as MFI.

4.9. Detection of ROS Production

ROS production was determined after 1, 3, 6, 18 or 24 h of WE treatment by using the probe dihydrorhodamine (DHR, 10 μM) [55], which was added during the last 15 min of incubation. Hydrogen peroxide was used as positive control. Additionally, cells were pre- treated for 30 min with Rot (2 μM) or o-Phe (10 μM) and co-treated for 6 h with WE (0.8 mg/mL). In some experiments, cells were co-treated for 6 h with WE (0.8 mg/mL) plus NAC (10 mM). After three washing in PBS, cellular fluorescence was imaged using a Leica DMLB/DFC300F fluorescence microscope (Leica Microsystems, Wetzlar, Germany) equipped with an Olympus ColorviewIIIu CCD camera (Polyphoto, Milan, Italy). Fluorescence images (100 cells per sample from randomly selected fields) were digitally acquired and processed for fluorescence determination at the single cell level on a personal computer using the public domain program, Image J. Mean fluorescence values were determined by averaging the fluorescence of at least 100 cells/treatment condition/experiment.

4.10. Analysis of Calreticulin Translocation, Hsp-70 and Hsp-90 Expression, and ATP Release

After 6 or 24 h of treatment, cells were washed and incubated with phycoerythrin-labeled calreticulin antibody (1:100, Abcam, San Francisco, CA, USA) or isotope-matched negative control (isotypic mouse IgG1 K Alexa Fluor 488®) (eBioscience, San Diego, CA, USA).

To analyze Hsp-70 and Hsp-90 expression, cells were incubated with an anti-Hsp-70 or anti-Hsp-90 antibody (1:100, Abcam, for both antibodies) and, after washing, incubated with fluorescein isothyocianate-labeled secondary antibody (1:100, Sigma, Merck Millipore, Darmstadt, Germany) or the isotype control. Mean fluorescence was detected via flow cytometry.

The kit ATPLite™ 1step (Perkin Elmer, Waltham, MA, USA) was used for the detection of ATP extracellular concentration. Jurkat cells were seeded and treated with WE in Hank's Balanced Salt Solution (HBSS) or complete medium for 6 and 24 h, respectively. At the end of incubation, supernatants were collected and treated with 100 μL of ATPLite 1step reagent containing luciferase and D-luciferin. After shaking for 2 min at 700 rpm using the orbital microplate shaker 711/+

(Asal srl, Florence, Italy), luminescence of the samples was measured in a 96-well black plate using the microplate reader Victor X3 (Perkin Elmer).

4.11. DNA Damage Analysis

Phosphorylation of histone P-H2A.X was used as marker of WE genotoxic potential. After, 6 h of treatment with WE, cells were fixed, permeabilized and incubated for 30 min in the dark at room temperature with an anti-P-H2A.X-Alexa Fluor® antibody (Merck Millipore). Etoposide 10 µM was used as positive control. Samples were analyzed via flow cytometry.

4.12. Flow Cytometry

EasyCyte 5HT (Merck Millipore) was used to perform all flow cytometric analyses, with the exception of the measurements of $[Ca^{2+}]_i$ performed by using a FACSCanto II (BD Bioscience, Franklin Lakes, NJ, USA). For each sample, approximately 5000 events were evaluated.

4.13. Statistical Analysis

All results are expressed as mean ± SEM of at least three independent experiments. Differences between treatments were assessed by t test or one-way ANOVA and Dunnet or Bonferroni was used as post-tests. All statistical analyses were performed using GraphPad InStat 5.0 version (GraphPad Prism, San Diego, CA, USA, 2007). $p < 0.05$ was considered significant.

Acknowledgments: This work was supported by FARB—Finanziamenti di Ateneo alla Ricerca di base 2012 (RFBO124222).

Author Contributions: C.F. and P.S. conceived and designed the experiments; E.T., C.C., E.C., E.d.G., M.T., A.R.D., and B.C. performed the experiments; C.F., E.T., C.C., S.P., A.G., and M.T. analyzed the data; and C.F., E.T., C.C., P.S., G.V., A.G., and P.H. wrote the paper.

Conflicts of Interest: The authors declare no conflict of interest.

References

1. Global Status Report on Noncommunicable Diseases 2014. Available online: http://www.who.int/nmh/publications/ncd-status-report-2014/en/ (accessed on 22 December 2015).
2. Fimognari, C.; Ferruzzi, L.; Turrini, E.; Carulli, G.; Lenzi, M.; Hrelia, P.; Cantelli-Forti, G. Metabolic and toxicological considerations of botanicals in anticancer therapy. *Expert Opin. Drug Metab. Toxicol.* **2012**, *8*, 819–832. [CrossRef] [PubMed]
3. Rivera, E.; Gomez, H. Chemotherapy resistance in metastatic breast cancer: The evolving role of ixabepilone. *Breast Cancer Res.* **2010**, *12* (Suppl. 2), 1–12. [CrossRef] [PubMed]
4. Kroemer, G.; Galluzzi, L.; Kepp, O.; Zitvogel, L. Immunogenic cell death in cancer therapy. *Annu. Rev. Immunol.* **2013**, *31*, 51–72. [CrossRef] [PubMed]
5. Hou, W.; Zhang, Q.; Yan, Z.; Chen, R.; Zeh, H.J., III; Kang, R.; Lotze, M.T.; Tang, D. Strange attractors: DAMPs and autophagy link tumor cell death and immunity. *Cell Death Dis.* **2013**, *4*, e966. [CrossRef] [PubMed]
6. Newman, D.J.; Cragg, G.M.; Snader, K.M. Natural products as sources of new drugs over the period 1981–2002. *J. Nat. Prod.* **2003**, *66*, 1022–1037. [CrossRef] [PubMed]
7. Chen, S.T.; Dou, J.; Temple, R.; Agarwal, R.; Wu, K.M.; Walker, S. New therapies from old medicines. *Nat. Biotechnol.* **2008**, *26*, 1077–1083. [CrossRef] [PubMed]
8. Newman, D.J.; Cragg, G.M. Natural products as sources of new drugs over the 30 years from 1981 to 2010. *J. Nat. Prod.* **2012**, *75*, 311–335. [CrossRef] [PubMed]
9. Nance, C.L. Clinical efficacy trials with natural products and herbal medicines. In *Phytotherapies: Efficacy, Safety and Regulation*; Ramzan, I., Ed.; John Wiley & Sons Inc.: Hoboken, NJ, USA, 2015; pp. 65–88.
10. Shanafelt, T.D.; Call, T.G.; Zent, C.S.; Leis, J.F.; LaPlant, B.; Bowen, D.A.; Roos, M.; Laumann, K.; Ghosh, A.K.; Lesnick, C.; et al. Phase 2 trial of daily, oral Polyphenon E in patients with asymptomatic, Rai stage 0 to II chronic lymphocytic leukemia. *Cancer* **2013**, *119*, 363–370. [CrossRef] [PubMed]

11. Kulkarni, S.K.; Dhir, A. Withania somnifera: An Indian ginseng. *Prog. Neuropsychopharmacol. Biol. Psychiatry* **2008**, *32*, 1093–1105. [CrossRef] [PubMed]

12. Masevhe, N.A.; McGaw, L.J.; Eloff, J.N. The traditional use of plants to manage candidiasis and related infections in Venda, South Africa. *J. Ethnopharmacol.* **2015**, *168*, 364–372. [CrossRef] [PubMed]

13. Khanna, D.; Sethi, G.; Ahn, K.S.; Pandey, M.K.; Kunnumakkara, A.B.; Sung, B.; Aggarwal, A.; Aggarwal, B.B. Natural products as a gold mine for arthritis treatment. *Curr. Opin. Pharmacol.* **2007**, *7*, 344–351. [CrossRef] [PubMed]

14. Nosalova, G.; Fleskova, D.; Jurecek, L.; Sadlonova, V.; Ray, B. Herbal polysaccharides and cough reflex. *Respir. Physiol. Neurobiol.* **2013**, *187*, 47–51. [CrossRef] [PubMed]

15. Jain, S.; Shukla, S.D.; Sharma, K.; Bhatnagar, M. Neuroprotective effects of *Withania somnifera* Dunn. in hippocampal sub-regions of female albino rat. *Phytother. Res.* **2001**, *15*, 544–548. [CrossRef] [PubMed]

16. Rai, M.; Jogee, P.S.; Agarkar, G.; Santos, C.A. Anticancer activities of *Withania somnifera*: Current research, formulations, and future perspectives. *Pharm. Biol.* **2016**, *54*, 189–197. [CrossRef] [PubMed]

17. Chaurasiya, N.D.; Uniyal, G.C.; Lal, P.; Misra, L.; Sangwan, N.S.; Tuli, R.; Sangwan, R.S. Analysis of withanolides in root and leaf of Withania somnifera by HPLC with photodiode array and evaporative light scattering detection. *Phytochem. Anal.* **2008**, *19*, 148–154. [CrossRef] [PubMed]

18. Sharma, A.; Singh, K.; Almasan, A. Histone H2AX phosphorylation: A marker for DNA damage. *Methods Mol. Biol.* **2012**, *920*, 613–626. [PubMed]

19. Vyas, A.R.; Singh, S.V. Molecular targets and mechanisms of cancer prevention and treatment by withaferin a, a naturally occurring steroidal lactone. *AAPS. J.* **2014**, *16*, 1–10. [CrossRef] [PubMed]

20. Choudharymy, M.I.; Yousuf, S.; Atta-Ur-Rahman. Withanolides: Chemistry and antitumor activity. In *Natural Products*; Ramawat, K.G., Merillon, J.M., Eds.; Springer-Verlag: Berlin, Germany; Heidelberg, Germany, 2013; pp. 3465–3495.

21. Ichikawa, H.; Takada, Y.; Shishodia, S.; Jayaprakasam, B.; Nair, M.G.; Aggarwal, B.B. Withanolides potentiate apoptosis, inhibit invasion, and abolish osteoclastogenesis through suppression of nuclear factor-kappaB (NF-kappaB) activation and NF-kappaB-regulated gene expression. *Mol. Cancer Ther.* **2006**, *5*, 1434–1445. [CrossRef] [PubMed]

22. Stewart, Z.A.; Westfall, M.D.; Pietenpol, J.A. Cell-cycle dysregulation and anticancer therapy. *Trends Pharmacol. Sci.* **2003**, *24*, 139–145. [CrossRef]

23. Shapiro, G.I.; Harper, J.W. Anticancer drug targets: Cell cycle and checkpoint control. *J. Clin. Investig.* **1999**, *104*, 1645–1653. [CrossRef] [PubMed]

24. Oh, J.H.; Lee, T.J.; Kim, S.H.; Choi, Y.H.; Lee, S.H.; Lee, J.M.; Kim, Y.H.; Park, J.W.; Kwon, T.K. Induction of apoptosis by withaferin A in human leukemia U937 cells through down-regulation of Akt phosphorylation. *Apoptosis* **2008**, *13*, 1494–1504. [CrossRef] [PubMed]

25. Senthil, V.; Ramadevi, S.; Venkatakrishnan, V.; Giridharan, P.; Lakshmi, B.S.; Vishwakarma, R.A.; Balakrishnan, A. Withanolide induces apoptosis in HL-60 leukemia cells via mitochondria mediated cytochrome c release and caspase activation. *Chem. Biol. Interact.* **2007**, *167*, 19–30. [CrossRef] [PubMed]

26. Kruman, I.; Guo, Q.; Mattson, M.P. Calcium and reactive oxygen species mediate staurosporine-induced mitochondrial dysfunction and apoptosis in PC12 cells. *J. Neurosci. Res.* **1998**, *51*, 293–308. [CrossRef]

27. Nicotera, P.; Orrenius, S. The role of calcium in apoptosis. *Cell Calcium* **1998**, *23*, 173–180. [CrossRef]

28. Salvioli, S.; Ardizzoni, A.; Franceschi, C.; Cossarizza, A. JC-1, but not DiOC6(3) or rhodamine 123, is a reliable fluorescent probe to assess $\Delta\Psi$ changes in intact cells: Implications for studies on mitochondrial functionality during apoptosis. *FEBS Lett.* **1997**, *411*, 77–82. [CrossRef]

29. Oza, V.P.; Parmar, P.P.; Kumar, S.; Subramanian, R.B. Anticancer properties of highly purified L-asparaginase from *Withania somnifera* L. against acute lymphoblastic leukemia. *Appl. Biochem. Biotechnol.* **2010**, *160*, 1833–1840. [CrossRef] [PubMed]

30. Hahm, E.R.; Moura, M.B.; Kelley, E.E.; Van, H.B.; Shiva, S.; Singh, S.V. Withaferin A-induced apoptosis in human breast cancer cells is mediated by reactive oxygen species. *PLoS ONE* **2011**, *6*, e23354. [CrossRef] [PubMed]

31. Li, X.; Zhu, F.; Jiang, J.; Sun, C.; Wang, X.; Shen, M.; Tian, R.; Shi, C.; Xu, M.; Peng, F.; *et al.* Synergistic antitumor activity of withaferin A combined with oxaliplatin triggers reactive oxygen species-mediated inactivation of the PI3K/AKT pathway in human pancreatic cancer cells. *Cancer Lett.* **2015**, *357*, 219–230. [CrossRef] [PubMed]

32. Malik, F.; Kumar, A.; Bhushan, S.; Khan, S.; Bhatia, A.; Suri, K.A.; Qazi, G.N.; Singh, J. Reactive oxygen species generation and mitochondrial dysfunction in the apoptotic cell death of human myeloid leukemia HL-60 cells by a dietary compound withaferin A with concomitant protection by N-acetyl cysteine. *Apoptosis* **2007**, *12*, 2115–2133. [CrossRef] [PubMed]

33. Mayola, E.; Gallerne, C.; Esposti, D.D.; Martel, C.; Pervaiz, S.; Larue, L.; Debuire, B.; Lemoine, A.; Brenner, C.; Lemaire, C. Withaferin A induces apoptosis in human melanoma cells through generation of reactive oxygen species and down-regulation of Bcl-2. *Apoptosis* **2011**, *16*, 1014–1027. [CrossRef] [PubMed]

34. Nishikawa, Y.; Okuzaki, D.; Fukushima, K.; Mukai, S.; Ohno, S.; Ozaki, Y.; Yabuta, N.; Nojima, H. Withaferin A Induces Cell Death Selectively in Androgen-Independent Prostate Cancer Cells but Not in Normal Fibroblast Cells. *PLoS ONE* **2015**, *10*, e0134137. [CrossRef] [PubMed]

35. Trachootham, D.; Alexandre, J.; Huang, P. Targeting cancer cells by ROS-mediated mechanisms: A radical therapeutic approach? *Nat. Rev. Drug Discov.* **2009**, *8*, 579–591. [CrossRef] [PubMed]

36. Kakar, S.S.; Jala, V.R.; Fong, M.Y. Synergistic cytotoxic action of cisplatin and withaferin A on ovarian cancer cell lines. *Biochem. Biophys. Res. Commun.* **2012**, *423*, 819–825. [CrossRef] [PubMed]

37. Yu, S.M.; Kim, S.J. Production of reactive oxygen species by withaferin A causes loss of type collagen expression and COX-2 expression through the PI3K/Akt, p38, and JNK pathways in rabbit articular chondrocytes. *Exp. Cell Res.* **2013**, *319*, 2822–2834. [CrossRef] [PubMed]

38. Yu, S.M.; Kim, S.J. Withaferin A-caused production of intracellular reactive oxygen species modulates apoptosis via PI3K/Akt and JNKinase in rabbit articular chondrocytes. *J. Korean Med. Sci.* **2014**, *29*, 1042–1053. [CrossRef] [PubMed]

39. Sestili, P.; Diamantini, G.; Bedini, A.; Cerioni, L.; Tommasini, I.; Tarzia, G.; Cantoni, O. Plant-derived phenolic compounds prevent the DNA single-strand breakage and cytotoxicity induced by tert-butylhydroperoxide via an iron-chelating mechanism. *Biochem. J.* **2002**, *364*, 121–128. [CrossRef] [PubMed]

40. Teeter, M.E.; Baginsky, M.L.; Hatefi, Y. Ectopic inhibition of the complexes of the electron transport system by rotenone, piericidin A, demerol and antimycin A. *Biochim. Biophys. Acta* **1969**, *172*, 331–333. [CrossRef]

41. Krysko, D.V.; Garg, A.D.; Kaczmarek, A.; Krysko, O.; Agostinis, P.; Vandenabeele, P. Immunogenic cell death and DAMPs in cancer therapy. *Nat. Rev. Cancer* **2012**, *12*, 860–875. [CrossRef] [PubMed]

42. Rao, R.V.; Ellerby, H.M.; Bredesen, D.E. Coupling endoplasmic reticulum stress to the cell death program. *Cell Death Differ.* **2004**, *11*, 372–380. [CrossRef] [PubMed]

43. Garg, A.D.; Nowis, D.; Golab, J.; Vandenabeele, P.; Krysko, D.V.; Agostinis, P. Immunogenic cell death, DAMPs and anticancer therapeutics: An emerging amalgamation. *Biochim. Biophys. Acta* **2010**, *1805*, 53–71. [CrossRef] [PubMed]

44. McKenna, M.K.; Gachuki, B.W.; Alhakeem, S.S.; Oben, K.N.; Rangnekar, V.M.; Gupta, R.C.; Bondada, S. Anti-cancer activity of withaferin A in B-cell lymphoma. *Cancer Biol. Ther.* **2015**, *16*, 1088–1098. [CrossRef] [PubMed]

45. Yu, Y.; Hamza, A.; Zhang, T.; Gu, M.; Zou, P.; Newman, B.; Li, Y.; Gunatilaka, A.A.; Zhan, C.G.; Sun, D. Withaferin A targets heat shock protein 90 in pancreatic cancer cells. *Biochem. Pharmacol.* **2010**, *79*, 542–551. [CrossRef] [PubMed]

46. Grover, A.; Shandilya, A.; Agrawal, V.; Pratik, P.; Bhasme, D.; Bisaria, V.S.; Sundar, D. Hsp90/Cdc37 chaperone/co-chaperone complex, a novel junction anticancer target elucidated by the mode of action of herbal drug Withaferin A. *BMC Bioinf.* **2011**, *12*. [CrossRef] [PubMed]

47. Malik, F.; Singh, J.; Khajuria, A.; Suri, K.A.; Satti, N.K.; Singh, S.; Kaul, M.K.; Kumar, A.; Bhatia, A.; Qazi, G.N. A standardized root extract of *Withania somnifera* and its major constituent withanolide-A elicit humoral and cell-mediated immune responses by up regulation of Th1-dominant polarization in BALB/c mice. *Life Sci.* **2007**, *80*, 1525–1538. [CrossRef] [PubMed]

48. Panjamurthy, K.; Manoharan, S.; Balakrishnan, S.; Suresh, K.; Nirmal, M.R.; Senthil, N.; Alias, L.M. Protective effect of Withaferin-A on micronucleus frequency and detoxication agents during experimental oral carcinogenesis. *Afr. J. Tradit. Complement. Altern. Med.* **2008**, *6*, 1–8. [CrossRef] [PubMed]

49. Panjamurthy, K.; Manoharan, S.; Menon, V.P.; Nirmal, M.R.; Senthil, N. Protective role of withaferin-A on 7,12-dimethylbenz(a)anthracene-induced genotoxicity in bone marrow of Syrian golden hamsters. *J. Biochem. Mol. Toxicol.* **2008**, *22*, 251–258. [CrossRef] [PubMed]

50. Cooley, K.; Szczurko, O.; Perri, D.; Mills, E.J.; Bernhardt, B.; Zhou, Q.; Seely, D. Naturopathic care for anxiety: A randomized controlled trial ISRCTN78958974. *PLoS ONE* **2009**, *4*, e6628. [CrossRef] [PubMed]

51. Ahmad, M.K.; Mahdi, A.A.; Shukla, K.K.; Islam, N.; Rajender, S.; Madhukar, D.; Shankhwar, S.N.; Ahmad, S. *Withania somnifera* improves semen quality by regulating reproductive hormone levels and oxidative stress in seminal plasma of infertile males. *Fertil. Steril.* **2010**, *94*, 989–996. [CrossRef] [PubMed]

52. Chopra, A.; Lavin, P.; Patwardhan, B.; Chitre, D. A 32-week randomized, placebo-controlled clinical evaluation of RA-11, an Ayurvedic drug, on osteoarthritis of the knees. *J. Clin. Rheumatol.* **2004**, *10*, 236–245. [CrossRef] [PubMed]

53. Wendt, E.R.; Ferry, H.; Greaves, D.R.; Keshav, S. Ratiometric analysis of fura red by flow cytometry: A technique for monitoring intracellular calcium flux in primary cell subsets. *PLoS ONE* **2015**, *10*, e0119532. [CrossRef] [PubMed]

54. Verriere, V.; Higgins, G.; Al-Alawi, M.; Costello, R.W.; McNally, P.; Chiron, R.; Harvey, B.J.; Urbach, V. Lipoxin A4 stimulates calcium-activated chloride currents and increases airway surface liquid height in normal and cystic fibrosis airway epithelia. *PLoS ONE* **2012**, *7*, e37746. [CrossRef] [PubMed]

55. Royall, J.A.; Ischiropoulos, H. Evaluation of 2′,7′-dichlorofluorescin and dihydrorhodamine 123 as fluorescent probes for intracellular H_2O_2 in cultured endothelial cells. *Arch. Biochem. Biophys.* **1993**, *302*, 348–355. [CrossRef] [PubMed]

MDPI

Communication

Cells Deficient in the Fanconi Anemia Protein FANCD2 are Hypersensitive to the Cytotoxicity and DNA Damage Induced by Coffee and Caffeic Acid

Estefanía Burgos-Morón [1], José Manuel Calderón-Montaño [1,2], Manuel Luis Orta [3], Emilio Guillén-Mancina [1], Santiago Mateos [3] and Miguel López-Lázaro [1,*]

[1] Department of Pharmacology, Faculty of Pharmacy, University of Seville, Profesor García González 2, 41012 Seville, Spain; eburgos1@us.es (E.B.-M.); jose.calderon@cabimer.es (J.M.C.-M.); eguillen@us.es (E.G.-M.)

[2] Department of Molecular Biology, Centro Andaluz de Biología Molecular y Medicina Regenerativa, University of Seville, Avda. Americo Vespucio s/n., 41092 Seville, Spain

[3] Department of Cell Biology, Faculty of Biology, University of Seville, Avda. Reina Mercedes s/n., 41012 Seville, Spain; morta2@us.es (M.L.O.); smateos@us.es (S.M.)

* Correspondence: mlopezlazaro@us.es; Tel.: +34-954-556-348; Fax: +34-954-233-765

Academic Editor: Carmela Fimognari
Received: 15 March 2016; Accepted: 1 July 2016; Published: 8 July 2016

Abstract: Epidemiological studies have found a positive association between coffee consumption and a lower risk of cardiovascular disorders, some cancers, diabetes, Parkinson and Alzheimer disease. Coffee consumption, however, has also been linked to an increased risk of developing some types of cancer, including bladder cancer in adults and leukemia in children of mothers who drink coffee during pregnancy. Since cancer is driven by the accumulation of DNA alterations, the ability of the coffee constituent caffeic acid to induce DNA damage in cells may play a role in the carcinogenic potential of this beverage. This carcinogenic potential may be exacerbated in cells with DNA repair defects. People with the genetic disease Fanconi Anemia have DNA repair deficiencies and are predisposed to several cancers, particularly acute myeloid leukemia. Defects in the DNA repair protein Fanconi Anemia D2 (FANCD2) also play an important role in the development of a variety of cancers (e.g., bladder cancer) in people without this genetic disease. This communication shows that cells deficient in FANCD2 are hypersensitive to the cytotoxicity (clonogenic assay) and DNA damage (γ-H2AX and 53BP1 focus assay) induced by caffeic acid and by a commercial lyophilized coffee extract. These data suggest that people with Fanconi Anemia, or healthy people who develop sporadic mutations in FANCD2, may be hypersensitive to the carcinogenic activity of coffee.

Keywords: coffee; caffeic acid; cancer; DNA damage; carcinogenesis; FANCD2; Fanconi anemia

1. Introduction

Coffee, one of the most widely consumed beverages in the world, can affect human health. Observational studies suggest that coffee consumption may lower the risk of developing diabetes, cardiovascular disorders, cirrhosis, and degenerative disorders such as Parkinson and Alzheimer disease [1,2]. Coffee consumption has also been associated with a decreased risk of total mortality; coffee drinkers had a lower risk of death from heart disease, chronic respiratory diseases, diabetes, pneumonia and influenza, and intentional self-harm [3]. No significant association between coffee consumption and total cancer mortality was found, however [3]. The effect of coffee consumption on the risk of cancer is inconclusive; some studies indicate that coffee may reduce the risk of some types of cancers [4–6], while others suggest that it may increase the risk of developing the disease [6,7]. According to the International Agency for Research on Cancer (IARC), coffee is classified as possibly carcinogenic to the human urinary bladder (IARC, Vol. 51). Several recent epidemiological studies have also revealed that maternal consumption of coffee during pregnancy may be associated with childhood leukemia [8–11]. A recent meta-analysis found that maternal coffee consumption during pregnancy significantly increased the risk of childhood acute lymphoblastic leukemia (ALL) and acute myeloid leukemia (AML) in a dose-response manner [11].

Caffeic acid is present in coffee in low amounts, but it is an important metabolite found in plasma and urine after coffee consumption [1,12]. In coffee, caffeic acid is typically bound to quinic acid to form an ester called 5-caffeoylquinic acid or chlorogenic acid [13]. Some studies have shown that caffeic acid possesses antioxidant and anti-genotoxic activities [14–19]. However, other investigations have revealed that this polyphenol generates reactive oxygen species (ROS) and produces carcinogenic effects [20–26]. In vivo studies have also found that diets containing caffeic acid induced tumors in animals [2]. Caffeic acid is therefore classified as possibly carcinogenic to humans (IARC, Vol. 56). This carcinogenic activity may be due to its ability to induce DNA damage [21–23], probably through a pro-oxidant mechanism [22,24–26].

Cancer is a disease caused by the accumulation of DNA alterations in our cells [27–30]. When our cells suffer DNA alterations, the DNA damage response machinery activates a variety of mechanisms to repair the damage [3]. These mechanisms are necessary for maintaining the integrity of the DNA, and therefore provide a fundamental biological barrier against carcinogenesis. Mutations in DNA repair genes result in the accumulation of cellular DNA damage and in predisposition to cancer. Some people are born with defects in genes involved in the DNA damage response machinery. People without germline mutations in DNA repair genes can also acquire them during a possible carcinogenic process. People with these mutations are particularly sensitive to the carcinogenic activity of compounds that induce types of DNA damage requiring these genes for repair.

The Fanconi anemia (FA) pathway plays an important role in the repair of several types of DNA damage, including interstrand crosslink, replication fork stalling and double strand breaks. The FA protein Fanconi Anemia D2 (FANCD2) is essential for proper functioning of this pathway; this protein is considered a surrogate marker for FA network activation. Mutations in a cluster of proteins of this pathway cause Fanconi Anemia, a rare autosomal recessive genetic disease characterized by bone marrow failure, congenital abnormalities, genomic instability and predisposition to several types of cancer, particularly acute myeloid leukemia [31–33]. Defects in proteins implicated in this DNA repair pathway has been described in several kinds of sporadic cancers, including bladder cancer and acute myeloid leukemia [34–37]. Deficiency in this DNA repair pathway makes cells hypersensitive to the cytotoxicity and DNA-damaging activities of a variety of agents, including ROS [38–40].

Since caffeic acid induces ROS-mediated DNA damage, and since the FA pathway participates in the repair of this type of DNA damage, we hypothesized that cells deficient in this pathway would be more susceptible than normal cells to the DNA-damaging effect of this dietary phytochemical. We report that human cells deficient in FANCD2 are hypersensitive to the cytotoxicity (clonogenic assay) and DNA damage (γ-H2AX and 53BP1 focus assay) induced by caffeic acid and by a commercial lyophilized coffee extract, and discuss the possible relevance of these results.

2. Results

2.1. Cells Deficient in FANCD2 Are Hypersensitive to the Cytotoxicity of Coffee and Caffeic Acid

Cells lacking the FA protein FANCD2 (PD20−/−) and cells complemented with FANCD2 (PD20+/+) were exposed for 4 h to caffeic acid and to a commercial lyophilized coffee extract. After 7 days of recovery in drug-free medium, cell survival was determined with the clonogenic assay. Figure 1 shows that the survival of cells lacking FANCD2 was significantly lower than that of proficient cells when exposed to several concentrations of caffeic acid (A) and coffee (B). This means that the cellular toxicity of coffee and caffeic acid is increased in cells lacking the DNA repair protein FANCD2.

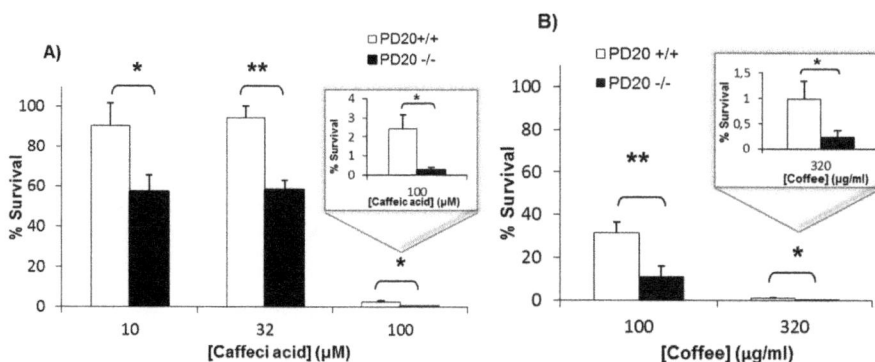

Figure 1. Cells deficient in Fanconi Anemia D2 (FANCD2) are hypersensitive to the cytotoxicity of caffeic acid (**A**) and a commercial lyophilized coffee extract (**B**). Parental PD20 cells with functional FANCD2 (PD20+/+) and PD20 cells lacking FANCD2 (PD20−/−) were treated with caffeic acid or coffee for 4 h. Then, the cells were allowed to form colonies in drug-free medium for 7 days, and the percentage of cell survival with respect to untreated cells was determined with the clonogenic assay. Data show the mean and standard deviation (SD) from at least 3 independent experiments. For statistical analysis, the *t*-test (paired, two-tailed) was used (* $p < 0.05$, ** $p < 0.01$).

2.2. Cells Deficient in FANCD2 Are Hypersensitive to the DNA Damage Induced by Coffee and Caffeic Acid

We used the immunofluorescence focus assay to measure the levels of DNA damage in FANCD2 deficient and proficient cells exposed to coffee and caffeic acid. We used specific antibodies to determine the levels of γ-H2AX and 53BP1 foci. An increase in the cellular levels of γ-H2AX foci is associated with the formation of double strand breaks (DSBs) in the DNA, but also with the formation of other types of DNA damage; γ-H2AX can therefore be considered as a marker of general DNA damage [4,41]. Formation of γ-H2AX foci is associated with recruitment of p53-binding protein 1 (53BP1), a regulator of the cellular response to DNA double-strand breaks. Therefore, the presence of 53BP1 foci is a specific marker of DSBs. Figure 2 shows that cells lacking FANCD2 developed higher levels of γ-H2AX foci and 53BP1 foci than non-deficient cells when exposed to coffee and caffeic acid.

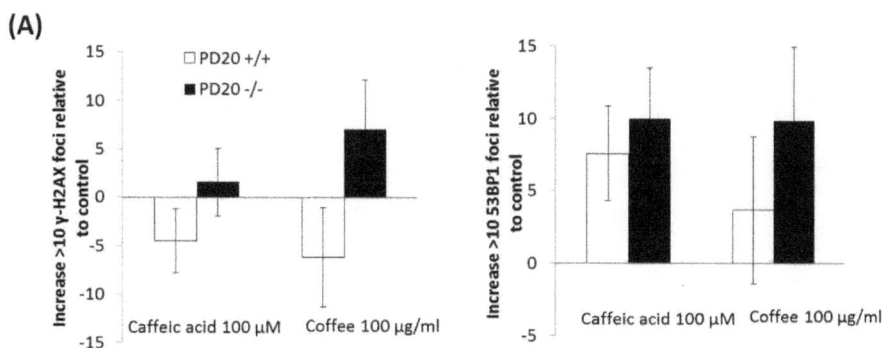

Figure 2. *Cont.*

(B)

(C)

(D)

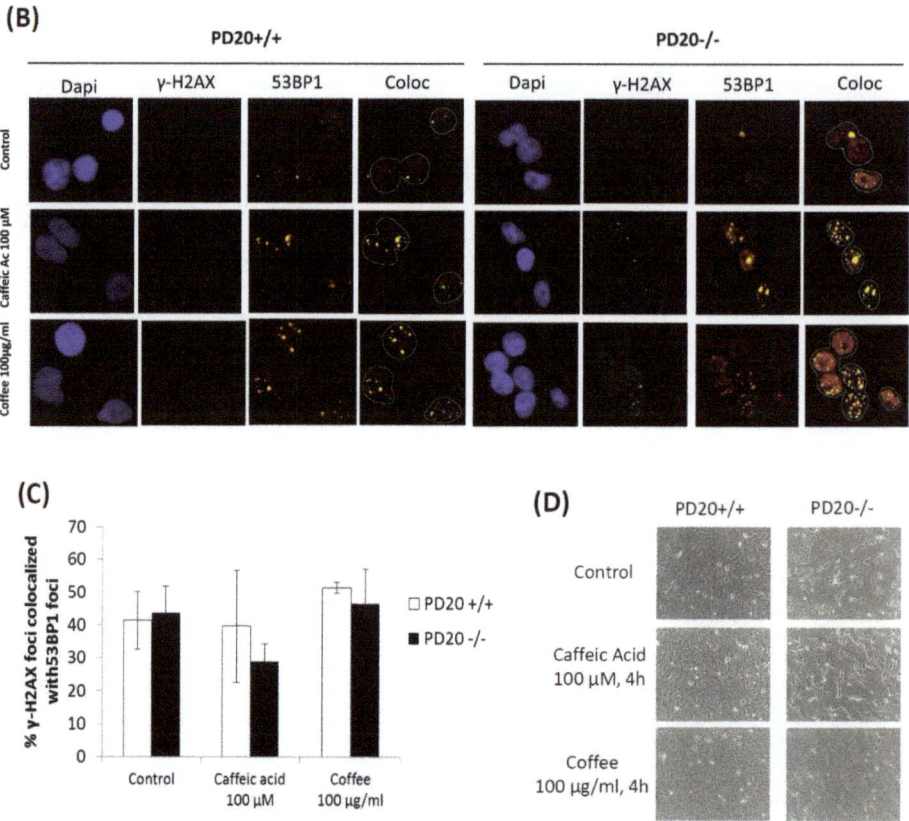

Figure 2. Cells deficient in FANCD2 (PD20−/−) are more sensitive than non-deficient cells (PD20+/+) to the DNA damage induced by a commercial lyophilized coffee extract and by caffeic acid. Cells were exposed for 4 h to caffeic acid 100 μM or coffee 100 μg/mL, and the levels of γ-H2AX and 53BP1 foci were measured with the Immunofluorescence focus assay. In (**A**), quantification of nuclear foci is presented. Data show the mean and standard deviation (SD) from at least 3 independent experiments; $p > 0.05$ (*t*-test, paired, two-tailed). Representative micrographs are shown in (**B**), where γ-H2AX foci appear as green spots, 53BP1 foci appear as orange spots and DAPI (4′,6-diamidino-2-phenylindole)-stained nucleus appear in blue. γ-H2AX foci colocalized with 53BP1 appear as yellow spots. Pictures were taken with an Olympus BX 61 microscope at 40-fold magnification (Figure 2B shows the part of the pictures that contained cells). In (**C**), the percentage of γ-H2AX foci colocalized with 53BP1 is represented. In (**D**), representative photographs of control cells and cells exposed for 4 h to caffeic acid 100 μM or coffee 100 μg/mL are shown.

3. Discussion

Coffee consumption may increase the risk of developing some types of cancer, including bladder cancer and childhood leukemia. Caffeic acid, a phenolic compound found in plasma and urine after coffee consumption, may contribute to the carcinogenic potential of coffee. Although the effect of coffee consumption on the risk of cancer is inconclusive, the International Agency for Research on Cancer has classified both coffee and caffeic acid as possibly carcinogenic to humans.

The ability of caffeic acid to induce DNA damage in cells [22–26] may play a key role in the carcinogenic potential of caffeic acid and coffee. It is well-known that dividing cells are more susceptible to DNA-damaging agents than non-dividing cells. Because cells divide actively during embryonic and fetal development, coffee consumption could be particularly carcinogenic during pregnancy. In addition, the carcinogenic potential of coffee would be higher in cells deficient in particular DNA repair proteins. Figure 1 shows that cells deficient in FANCD2, a critical DNA repair protein of the Fanconi Anemia pathway, are hypersensitive to the cytotoxicity of caffeic acid and coffee. This suggests that coffee and caffeic acid cause more DNA damage in cells deficient in this DNA repair protein. Figure 2 shows that cells lacking FANCD2 developed higher levels γ-H2AX foci than non-deficient cells when exposed to coffee and caffeic acid. This suggests that the DNA-damaging effects of coffee and caffeic acid are increased in cells with defects in the DNA repair protein FANCD2. The levels of 53BP1 foci were also increased in cells lacking FANCD2, therefore indicating that coffee and caffeic acid induce more DSBs in cells deficient in this DNA repair protein. Together, this data indicate that cells deficient in FANCD2 are hypersensitive to the cytotoxicity and DNA-damaging activities of caffeic acid and coffee, and suggest that people with mutations in FANCD2 may be hypersensitive to their carcinogenic activity.

The bioavailability of caffeic acid in humans is relatively high. The plasma and urinary levels of caffeic acid in humans after coffee consumption are typically in the nanomolar and low micromolar range [12,42,43]. For example, a study showed that the peak concentration of caffeic acid in the plasma of ten healthy adults was 1.1 ± 0.9 μM; its concentration in urine was highly variable, ranging from 0.07 to 9.43 μM [5]. Figure 2 shows that the concentration of caffeic acid required to detect DNA damage in cells is high (100 μM), probably because the sensitivity of the immunofluorescence focus assay is relatively low. But Figure 1 shows that, when treated with caffeic acid 10 μM, the survival of cells deficient in the DNA repair protein FANCD2 is approximately 60% of that of untreated cells. This suggests that caffeic acid 10 μM induces cytotoxic levels of DNA damage, and that non-cytotoxic levels of DNA damage probably occur at lower concentrations; these concentrations are possibly similar to those achieved in plasma and urine after coffee consumption.

The high concentrations of caffeic acid detected in the urine of some healthy volunteers after coffee consumption [6] indicate that this phytochemical can be accumulated in the urinary bladder. This may explain why coffee is classified as possibly carcinogenic to the human urinary bladder (IARC, Vol. 51). FANCD2 deficiencies have been described in bladder cancer [34–36]. Together, this supports the idea that coffee and caffeic acid may increase the risk of bladder cancer, particularly in people with germline or sporadic mutations in the DNA repair protein FANCD2.

The carcinogenic activity of coffee may be mediated not only by the pro-oxidant activity of caffeic acid [22,24–26], but also by other pro-oxidant coffee constituents. Several coffee constituents, including chlorogenic acid and hydroquinone, are known to generate hydrogen peroxide [44–47], and hydrogen peroxide induces DNA damage and plays a key role in cancer development [48]. The cytotoxicity and DNA-damaging activities of hydrogen peroxide and other ROS are higher in cells deficient in DNA repair proteins of the Anemia Fanconi pathway [38–40]. This suggests that the carcinogenic potential of coffee is mediated by the ability of caffeic acid and other coffee constituents to induce pro-oxidant DNA damage, and that this carcinogenic potential is higher in cells lacking DNA repair proteins of the Anemia Fanconi pathway. It is important to clarify that some polyphenols can both prevent and induce oxidative DNA damage, mainly depending on their concentration. At low concentrations, they can reduce the levels of ROS and prevent DNA damage. At higher concentrations, however, some polyphenols (particularly those containing catechol or pyrogallol moieties in their structure, such as caffeic acid) can generate hydrogen peroxide by an autoxidation mechanism. This mechanism involves the oxidation of polyphenols to semiquinones in a process in which oxygen is reduced to superoxide anion and hydrogen peroxide. Through the Fenton reaction, hydrogen peroxide produces hydroxyl radicals that cause oxidative DNA damage [49,50].

In summary, although the effect of coffee consumption on the risk of cancer is inconclusive, some studies indicate that coffee may increase the risk of developing some cancers [6,7]. Here we report that cells deficient in the DNA repair protein FANCD2 are hypersensitive to the cytotoxic and

DNA-damaging activities of a commercial lyophilized coffee extract and caffeic acid (an important metabolite found in plasma and urine after coffee consumption). These data suggest that people with Fanconi Anemia, or healthy people who develop sporadic mutations in FANCD2, may be hypersensitive to the carcinogenic activity of coffee.

4. Materials and Methods

4.1. Chemicals and Cell Lines

Caffeic acid was purchased from Sigma ($\geqslant$98.0%, Sigma-Aldrich, St. Louis, MO, USA). A commercial lyophilized coffee extract (NESCAFÉ® Classic, Barcelona, Spain) was used. In all experiments, caffeic acid and the coffee extract were tested individually, not mixed together. The human Fanconi deficient (PD20 FANCD2$-$/$-$) and proficient (PD20 FANCD2$-$/$-$ complemented with FANCD2) cells were kindly provided by Dr. Thomas Helleday and by Dr. Jordi Surrallés Calonge. Cells were maintained in DMEM supplemented with 2 mM glutamine, 50 µg/mL penicillin, 50 µg/mL streptomycin and 20% fetal bovine serum. Cells were cultured at 37 °C in a humidified atmosphere containing 5% CO_2.

4.2. Clonogenic Assay

Cell survival was measured with the clonogenic assay. Cells were plated at low density onto 6 cm Petri dishes. Cells were treated with caffeic acid or coffee for 4 h; then drugs were removed and fresh media was added to allow the cells to grow for 7 days. Colonies were stained with methylene blue prepared in methanol (4 g/L). Surviving colonies made up 50 cells per colony were counted and the data were corrected according to cloning efficiencies of control cells.

4.3. Immunofluorescence γ-H2AX and 53BP1 Focus Assay

To evaluate DNA damage, detection of γ-H2AX and 53BP1 foci were quantified by immunofluorescence using the focus assay. The γ-H2AX focus assay is based on the ability of double-strand breaks (DSBs) to trigger phosphorylation of histone H2AX on Ser-139, which leads to the formation of nuclear foci that can be visualized with anti-γ-H2AX antibodies. Following the induction of double strand breaks, formation of γ-H2AX foci is associated with recruitment of p53-binding protein 1 (53BP1). Accumulation of 53BP1 can also be visualized as foci with anti-53BP1 antibodies. After treatments, cells were washed three times with PBS, and incubated for 30 s with cold 0.1% Triton-X in PBS to pre-extract soluble proteins. Afterward, cells were fixed with 4% paraformaldehyde in PBS for 10 min at room temperature and washed three times with PBS. After fixation, cells were permeabilized with 0.5% Triton X-100 in PBS for 5 min and then blocked three times with 0.1% Tween 20, 1% BSA in PBS for 5 min each. Cells were then incubated for 1 h with a mouse anti-γ-H2AX monoclonal antibody (Upstate; 1:800 dilution). Cells were washed three times with PBS and blocked three times prior to the incubation with a secondary anti-mouse antibody linked to Alexa Fluor 488 (Invitrogen; 1:500 dilution) for 1 h. Cells were washed with PBS, blocked, and incubated overnight with the rabbit anti-53BP1 primary antibody (Bethyl, 1:1000 dilution). After incubation, cells were washed with PBS, blocked and washed again with PBS as indicated before. DNA was stained with DAPI (4′,6-diamidino-2-phenylindole) and immunofluorescence was observed at 40-fold magnification with an Olympus BX 61 microscope. A total of at least 100 cells/dose were scored, and cells with 10 or more foci were scored as positive [44,51].

Author Contributions: S.M. and M.L-L. conceived and designed the experiments; E.B.-M., J.M.C.-M., M.L.O. and E.G.-M. performed the experiments and analyzed the data; E.B-M. and M.L.-L. wrote the paper.

Conflicts of Interest: The authors declare no conflict of interest.

Abbreviations

ALL	acute lymphoblastic leukemia
AML	acute myeloid leukemia
DSBs	double strand breaks
FA	Fanconi anemia
IARC	International Agency for Research on Cancer
ROS	reactive oxygen species

References

1. Ludwig, I.A.; Clifford, M.N.; Lean, M.E.; Ashihara, H.; Crozier, A. Coffee: Biochemistry and potential impact on health. *Food Funct.* **2014**, *5*, 1695–1717. [CrossRef] [PubMed]
2. Higdon, J.V.; Frei, B. Coffee and health: A review of recent human research. *Crit. Rev. Food Sci. Nutr.* **2006**, *46*, 101–123. [CrossRef] [PubMed]
3. Ding, M.; Satija, A.; Bhupathiraju, S.N.; Hu, Y.; Sun, Q.; Han, J.; Lopez-Garcia, E.; Willett, W.; van Dam, R.M.; Hu, F.B. Association of Coffee Consumption with Total and Cause-Specific Mortality in 3 Large Prospective Cohorts. *Circulation* **2015**, *132*, 2305–2315. [CrossRef] [PubMed]
4. Bohn, S.K.; Blomhoff, R.; Paur, I. Coffee and cancer risk, epidemiological evidence, and molecular mechanisms. *Mol. Nutr. Food Res.* **2014**, *58*, 915–930. [CrossRef] [PubMed]
5. Sugiyama, K.; Sugawara, Y.; Tomata, Y.; Nishino, Y.; Fukao, A.; Tsuji, I. The association between coffee consumption and bladder cancer incidence in a pooled analysis of the Miyagi Cohort Study and Ohsaki Cohort Study. *Eur. J. Cancer Prev.* **2016**. [CrossRef] [PubMed]
6. Arab, L. Epidemiologic evidence on coffee and cancer. *Nutr. Cancer* **2010**, *62*, 271–283. [CrossRef] [PubMed]
7. Wu, W.; Tong, Y.; Zhao, Q.; Yu, G.; Wei, X.; Lu, Q. Coffee consumption and bladder cancer: A meta-analysis of observational studies. *Sci. Rep.* **2015**, *5*, 9051. [CrossRef] [PubMed]
8. Bonaventure, A.; Rudant, J.; Goujon-Bellec, S.; Orsi, L.; Leverger, G.; Baruchel, A.; Bertrand, Y.; Nelken, B.; Pasquet, M.; Michel, G.; et al. Childhood acute leukemia, maternal beverage intake during pregnancy, and metabolic polymorphisms. *Cancer Causes Control* **2013**, *24*, 783–793. [CrossRef] [PubMed]
9. Thomopoulos, T.P.; Ntouvelis, E.; Diamantaras, A.A.; Tzanoudaki, M.; Baka, M.; Hatzipantelis, E.; Kourti, M.; Polychronopoulou, S.; Sidi, V.; Stiakaki, E.; et al. Maternal and childhood consumption of coffee, tea and cola beverages in association with childhood leukemia: A meta-analysis. *Cancer Epidemiol.* **2015**, *39*, 1047–1059. [CrossRef] [PubMed]
10. Cheng, J.; Su, H.; Zhu, R.; Wang, X.; Peng, M.; Song, J.; Fan, D. Maternal coffee consumption during pregnancy and risk of childhood acute leukemia: A metaanalysis. *Am. J. Obstet. Gynecol.* **2014**, *210*, 151. [CrossRef] [PubMed]
11. Orsi, L.; Rudant, J.; Ajrouche, R.; Leverger, G.; Baruchel, A.; Nelken, B.; Pasquet, M.; Michel, G.; Bertrand, Y.; Ducassou, S.; et al. Parental smoking, maternal alcohol, coffee and tea consumption during pregnancy, and childhood acute leukemia: The ESTELLE study. *Cancer Causes Control* **2015**, *26*, 1003–1017. [CrossRef] [PubMed]
12. Farah, A.; Monteiro, M.; Donangelo, C.M.; Lafay, S. Chlorogenic acids from green coffee extract are highly bioavailable in humans. *J. Nutr.* **2008**, *138*, 2309–2015. [CrossRef] [PubMed]
13. Clifford, M.N. Chlorogenic acids and other cinnamates – nature, occurrence and dietary burden. *J. Sci.Food Agric.* **1999**, *79*, 362–372. [CrossRef]
14. Sato, Y.; Itagaki, S.; Kurokawa, T.; Ogura, J.; Kobayashi, M.; Hirano, T.; Sugawara, M.; Iseki, K. In vitro and in vivo antioxidant properties of chlorogenic acid and caffeic acid. *Int. J. Pharm.* **2011**, *403*, 136–138. [CrossRef] [PubMed]
15. Kono, Y.; Kobayashi, K.; Tagawa, S.; Adachi, K.; Ueda, A.; Sawa, Y.; Shibata, H. Antioxidant activity of polyphenolics in diets. Rate constants of reactions of chlorogenic acid and caffeic acid with reactive species of oxygen and nitrogen. *Biochim. Biophys. Acta* **1997**, *1335*, 335–342. [CrossRef]
16. Liang, G.; Shi, B.; Luo, W.; Yang, J. The protective effect of caffeic acid on global cerebral ischemia-reperfusion injury in rats. *Behav. Brain Funct.* **2015**, *11*, 18. [CrossRef] [PubMed]

17. Rosendahl, A.H.; Perks, C.M.; Zeng, L.; Markkula, A.; Simonsson, M.; Rose, C.; Ingvar, C.; Holly, J.M.; Jernstrom, H. Caffeine and Caffeic Acid Inhibit Growth and Modify Estrogen Receptor and Insulin-like Growth Factor I Receptor Levels in Human Breast Cancer. *Clin. Cancer Res.* **2015**, *21*, 1877–1887. [CrossRef] [PubMed]

18. Bakuradze, T.; Lang, R.; Hofmann, T.; Schipp, D.; Galan, J.; Eisenbrand, G.; Richling, E. Coffee consumption rapidly reduces background DNA strand breaks in healthy humans: Results of a short term repeated uptake intervention study. *Mol. Nutr. Food Res.* **2016**, *60*, 682–686. [CrossRef] [PubMed]

19. Rehman, M.U.; Sultana, S. Attenuation of oxidative stress, inflammation and early markers of tumor promotion by caffeic acid in Fe-NTA exposed kidneys of Wistar rats. *Mol. Cell. Biochem.* **2011**, *357*, 115–124. [CrossRef] [PubMed]

20. Hagiwara, A.; Hirose, M.; Takahashi, S.; Ogawa, K.; Shirai, T.; Ito, N. Forestomach and kidney carcinogenicity of caffeic acid in F344 rats and C57BL/6N$_x$ C$_3$H/HeN F$_1$ mice. *Cancer Res.* **1991**, *51*, 5655–5660. [PubMed]

21. Szeto, Y.T.; Collins, A.R.; Benzie, I.F. Effects of dietary antioxidants on DNA damage in lysed cells using a modified comet assay procedure. *Mutat. Res.* **2002**, *500*, 31–38. [CrossRef]

22. Bhat, S.H.; Azmi, A.S.; Hadi, S.M. Prooxidant DNA breakage induced by caffeic acid in human peripheral lymphocytes: Involvement of endogenous copper and a putative mechanism for anticancer properties. *Toxicol. Appl. Pharmacol.* **2007**, *218*, 249–255. [CrossRef] [PubMed]

23. Maistro, E.L.; Angeli, J.P.; Andrade, S.F.; Mantovani, M.S. In vitro genotoxicity assessment of caffeic, cinnamic and ferulic acids. *Genet. Mol. Res.* **2011**, *10*, 1130–1140. [CrossRef] [PubMed]

24. Inoue, S.; Ito, K.; Yamamoto, K.; Kawanishi, S. Caffeic acid causes metal-dependent damage to cellular and isolated DNA through H$_2$O$_2$ formation. *Carcinogenesis* **1992**, *13*, 1497–1502. [CrossRef] [PubMed]

25. Li, Y.; Trush, M.A. Reactive oxygen-dependent DNA damage resulting from the oxidation of phenolic compounds by a copper-redox cycle mechanism. *Cancer Res.* **1994**, *54*, 1895s–1898s. [PubMed]

26. Babich, H.; Schuck, A.G.; Weisburg, J.H.; Zuckerbraun, H.L. Research strategies in the study of the pro-oxidant nature of polyphenol nutraceuticals. *J. Toxicol.* **2011**, *2011*, 467305. [CrossRef] [PubMed]

27. Knudson, A.G. Cancer genetics. *Am. J. Med. Genet.* **2002**, *111*, 96–102. [CrossRef] [PubMed]

28. Greenman, C.; Stephens, P.; Smith, R.; Dalgliesh, G.L.; Hunter, C.; Bignell, G.; Davies, H.; Teague, J.; Butler, A.; Stevens, C.; et al. Patterns of somatic mutation in human cancer genomes. *Nature* **2007**, *446*, 153–158. [CrossRef] [PubMed]

29. Burrell, R.A.; McGranahan, N.; Bartek, J.; Swanton, C. The causes and consequences of genetic heterogeneity in cancer evolution. *Nature* **2013**, *501*, 338–345. [CrossRef] [PubMed]

30. Stratton, M.R.; Campbell, P.J.; Futreal, P.A. The cancer genome. *Nature* **2009**, *458*, 719–724. [CrossRef] [PubMed]

31. Errol, C.F.; Walker, G.C.; Siede, W. *DNA Repair and Mutagenesis*; ASM Press: Washington, DC, USA, 1995.

32. D'Andrea, A.D.; Grompe, M. Molecular biology of Fanconi anemia: Implications for diagnosis and therapy. *Blood* **1997**, *90*, 1725–1736. [PubMed]

33. Willers, H.; Kachnic, L.A.; Luo, C.M.; Li, L.; Purschke, M.; Borgmann, K.; Held, K.D.; Powell, S.N. Biomarkers and mechanisms of FANCD2 function. *J. Biomed. Biotechnol.* **2008**, *2008*, 821529. [CrossRef] [PubMed]

34. Panneerselvam, J.; Pickering, A.; Zhang, J.; Wang, H.; Tian, H.; Zheng, J.; Fei, P. A hidden role of the inactivated FANCD2: Upregulating DeltaNp63. *Oncotarget* **2013**, *4*, 1416–1426. [CrossRef] [PubMed]

35. Zhang, J.; Zhao, D.; Park, H.K.; Wang, H.; Dyer, R.B.; Liu, W.; Klee, G.G.; McNiven, M.A.; Tindall, D.J.; Molina, J.R.; et al. FAVL elevation in human tumors disrupts Fanconi anemia pathway signaling and promotes genomic instability and tumor growth. *J. Clin. Investig.* **2010**, *120*, 1524–1534. [CrossRef] [PubMed]

36. Panneerselvam, J.; Park, H.K.; Zhang, J.; Dudimah, F.D.; Zhang, P.; Wang, H.; Fei, P. FAVL impairment of the Fanconi anemia pathway promotes the development of human bladder cancer. *Cell Cycle* **2012**, *11*, 2947–2455. [CrossRef] [PubMed]

37. Tischkowitz, M.D.; Morgan, N.V.; Grimwade, D.; Eddy, C.; Ball, S.; Vorechovsky, I.; Langabeer, S.; Stoger, R.; Hodgson, S.V.; Mathew, C.G. Deletion and reduced expression of the Fanconi anemia FANCA gene in sporadic acute myeloid leukemia. *Leukemia* **2004**, *18*, 420–425. [CrossRef] [PubMed]

38. Joenje, H.; Arwert, F.; Eriksson, A.W.; de Koning, H.; Oostra, A.B. Oxygen-dependence of chromosomal aberrations in Fanconi's anaemia. *Nature* **1981**, *290*, 142–143. [CrossRef] [PubMed]

39. Saito, H.; Hammond, A.T.; Moses, R.E. Hypersensitivity to oxygen is a uniform and secondary defect in Fanconi anemia cells. *Mutat. Res.* **1993**, *294*, 255–262. [CrossRef]

40. Takeuchi, T.; Morimoto, K. Increased formation of 8-hydroxydeoxyguanosine, an oxidative DNA damage, in lymphoblasts from Fanconi's anemia patients due to possible catalase deficiency. *Carcinogenesis* **1993**, *14*, 1115–1120. [CrossRef] [PubMed]

41. Monteiro, F.L.; Baptista, T.; Amado, F.; Vitorino, R.; Jeronimo, C.; Helguero, L.A. Expression and functionality of histone H2A variants in cancer. *Oncotarget* **2014**, *5*, 3428–4343. [CrossRef] [PubMed]

42. Wittemer, S.M.; Ploch, M.; Windeck, T.; Muller, S.C.; Drewelow, B.; Derendorf, H.; Veit, M. Bioavailability and pharmacokinetics of caffeoylquinic acids and flavonoids after oral administration of Artichoke leaf extracts in humans. *Phytomedicine* **2005**, *12*, 28–38. [CrossRef] [PubMed]

43. Renouf, M.; Guy, P.A.; Marmet, C.; Fraering, A.L.; Longet, K.; Moulin, J.; Enslen, M.; Barron, D.; Dionisi, F.; Cavin, C.; et al. Measurement of caffeic and ferulic acid equivalents in plasma after coffee consumption: Small intestine and colon are key sites for coffee metabolism. *Mol. Nutr. Food Res.* **2010**, *54*, 760–766. [CrossRef] [PubMed]

44. Burgos-Moron, E.; Calderon-Montano, J.M.; Orta, M.L.; Pastor, N.; Perez-Guerrero, C.; Austin, C.; Mateos, S.; Lopez-Lazaro, M. The Coffee Constituent Chlorogenic Acid Induces Cellular DNA Damage and Formation of Topoisomerase I- and II-DNA Complexes in Cells. *J. Agric. Food Chem.* **2012**, *60*, 7384–7391. [CrossRef] [PubMed]

45. Long, L.H.; Halliwell, B. Coffee drinking increases levels of urinary hydrogen peroxide detected in healthy human volunteers. *Free Radic. Res.* **2000**, *32*, 463–467. [CrossRef] [PubMed]

46. Hiramoto, K.; Kida, T.; Kikugawa, K. Increased urinary hydrogen peroxide levels caused by coffee drinking. *Biol. Pharm. Bull.* **2002**, *25*, 1467–1471. [CrossRef] [PubMed]

47. Halliwell, B.; Long, L.H.; Yee, T.P.; Lim, S.; Kelly, R. Establishing biomarkers of oxidative stress: The measurement of hydrogen peroxide in human urine. *Curr. Med. Chem.* **2004**, *11*, 1085–1092. [CrossRef] [PubMed]

48. Lopez-Lazaro, M. Dual role of hydrogen peroxide in cancer: Possible relevance to cancer chemoprevention and therapy. *Cancer Lett.* **2007**, *252*, 1–8. [CrossRef] [PubMed]

49. Akagawa, M.; Shigemitsu, T.; Suyama, K. Production of hydrogen peroxide by polyphenols and polyphenol-rich beverages under quasi-physiological conditions. *Biosci. Biotechnol. Biochem.* **2003**, *67*, 2632–2640. [CrossRef] [PubMed]

50. Mochizuki, M.; Yamazaki, S.; Kano, K.; Ikeda, T. Kinetic analysis and mechanistic aspects of autoxidation of catechins. *Biochim. Biophys. Acta* **2002**, *1569*, 35–44. [CrossRef]

51. Orta, M.L.; Calderon-Montano, J.; Dominguez, I.; Pastor, N.; Burgos-Moron, E.; Lopez-Lazaro, M.; Cortes, F.; Mateos, S.; Helleday, T. 5-Aza-2′-deoxycytidine causes replication lesions that require Fanconi anemia-dependent homologous recombination for repair. *Nucleic Acids Res.* **2013**, *41*, 5827–5836. [CrossRef] [PubMed]

toxins

MDPI

Review

Roles of Dietary Phytoestrogens on the Regulation of Epithelial-Mesenchymal Transition in Diverse Cancer Metastasis

Geum-A. Lee, Kyung-A. Hwang * and Kyung-Chul Choi *

Laboratory of Biochemistry and Immunology, College of Veterinary Medicine, Chungbuk National University, Cheongju, Chungbuk 361-763, Korea; mmanuraa@gmail.com
* Correspondence: hka9400@naver.com (K.-A.H.); kchoi@cbu.ac.kr (K.-C.C.);
Tel.: +82-43-249-1745 (K.-A.H.); +82-43-261-3664 (K.-C.C.); Fax: +82-43-267-3150 (K.-A.H. & K.-C.C.)

Academic Editor: Carmela Fimognari
Received: 6 March 2016; Accepted: 19 May 2016; Published: 24 May 2016

Abstract: Epithelial-mesenchymal transition (EMT) plays a key role in tumor progression. The cells undergoing EMT upregulate the expression of cell motility-related proteins and show enhanced migration and invasion. The hallmarks of EMT in cancer cells include changed cell morphology and increased metastatic capabilities in cell migration and invasion. Therefore, prevention of EMT is an important tool for the inhibition of tumor metastasis. A novel preventive therapy is needed, such as treatment of natural dietary substances that are nontoxic to normal human cells, but effective in inhibiting cancer cells. Phytoestrogens, such as genistein, resveratrol, kaempferol and 3,3'-diindolylmethane (DIM), can be raised as possible candidates. They are plant-derived dietary estrogens, which are found in tea, vegetables and fruits, and are known to have various biological efficacies, including chemopreventive activity against cancers. Specifically, these phytoestrogens may induce not only anti-proliferation, apoptosis and cell cycle arrest, but also anti-metastasis by inhibiting the EMT process in various cancer cells. There have been several signaling pathways found to be associated with the induction of the EMT process in cancer cells. Phytoestrogens were demonstrated to have chemopreventive effects on cancer metastasis by inhibiting EMT-associated pathways, such as Notch-1 and TGF-beta signaling. As a result, phytoestrogens can inhibit or reverse the EMT process by upregulating the expression of epithelial phenotypes, including *E*-cadherin, and downregulating the expression of mesenchymal phenotypes, including *N*-cadherin, Snail, Slug, and vimentin. In this review, we focused on the important roles of phytoestrogens in inhibiting EMT in many types of cancer and suggested phytoestrogens as prominent alternative compounds to chemotherapy.

Keywords: dietary phytoestrogens; DIM; kaempferol; resveratrol; genistein; epithelial-mesenchymal transition; cancer metastasis

1. Introduction

Phytochemicals are chemical compounds that occur naturally in plants, amounting to as many as 4000 different chemicals. Some phytochemicals have the biological significance of inhibiting the invasion of different species of plants by acting as toxic compounds [1,2]. This characteristic of phytochemicals has been utilized in curing diverse human diseases. A vast array of plant-derived natural compounds has been reported to have substantial chemopreventive effects against cancer, in opposition to the health risk of environmental carcinogens [3]. At present, inhibiting human carcinogenesis using plant-derived compounds is considered as a vital and urgent challenge, despite some phytochemicals having been used for targeting many forms of cancer as major sources of highly effective conventional drugs [4–6].

Among diverse groups of phytochemicals, phytoestrogens, which are plant-derived xenoestrogens and mostly found in soy, vegetables and fruits, are considered as strong sources of cancer-preventive phytonutrition to inhibit the development and progression of many types of cancer [4,7].

As a kind of xenoestrogen, phytoestrogens specifically have distinctive cancer-preventive effects on estrogen-related cancers. In general, sex hormones, including estrogens, have been known to be closely linked to the pathogenesis of several types of cancer in the reproductive organs [8–10]. Many diseases, like breast, ovarian, endometrial and cervical cancers, have been called estrogen-receptor (ER)-positive cancers, because the actions of estrogen related to cancer biology are mediated via ERs, which comprise ERα and ERβ, mostly present in the nucleus [11,12]. Independently of endogenous estrogens, selective estrogen receptor modulators (SERMs), which are a class of drugs that act on ER, act with agonistic or antagonistic actions in several target tissues [13]. Phytoestrogens, which are also a kind of SERM, are known to bind ER with affinities at least 10,000-times lower than that of 17β-estradiol (E2) and also act as ER agonists or antagonists [14]. The chemopreventive effects of phytoestrogens are associated with their antagonistic effects on ER [15]. In addition to the actions of phytoestrogens via ERs, they could have protective effects on the initiation and progression of estrogen-related cancers by specifically inhibiting the circulating precursors of estrogens [16–20]. In addition, they can inhibit crucial steroidogenic enzyme activity, including the conversion of E2 from circulating hormones, such as androgens and estrogen sulfate [20]. In estrogen biosynthesis and metabolism, phytoestrogens have been shown to inhibit several crucial enzymes in aromatase pathway, such as 17β- and 3β-hydroxysteroid dehydrogenase (HSD), which catalyze the dehydrogenation of 17-hydroxysteroids in steroidogenesis and control the interconversion of androstenedione and testosterone, and E2 and estrone, respectively [21,22].

Phytoestrogens are generally classified into four main classes: isoflavones (genistein, daidzein, kaempferol), lignans (secoisolariciresinol, matairesinol, pinoresinol, lariciresinol), coumestan (coumestrol) and stilbenes (resveratrol) [23,24]. Western populations have been known to intake more foods containing lignans, while Asian populations eat more soy foods containing isoflavones [25]. Lignans are included in diverse groups of non-flavonoid compounds widely distributed in whole grain cereals, beans, berries, nuts and various seeds [26,27]. A wealth of lignans exists as secoisolariciresinol, matairesinol, lariciresinol and pinoresinol, and they are converted into enterolignans by the intestinal microbiota to be absorbed into the human body [27,28]. Tea, fruits, vegetables and grains account for over 85% of the daily intake of lignans, such as matairesinol and secoisolariciresinol [29]. Isoflavones are naturally-occurring phenolic flavonoid compounds, known to act as phytoestrogens in mammals. Soybeans are the most common source of isoflavones among vegetables; the major isoflavones in soybean are genistein and daidzein, which are well-known phytoestrogens [30]. Coumestans occur mainly in bean sprouts during germination, and the main compound in this subgroup is coumestrol, mostly found in peas and beans [23]. Stilbenoids are hydroxylated derivatives of stilbene and belong to the family of phenylpropanoids [31]. Resveratrol is a typical stilbenoid, which is mainly found in grapes and wines [32].

In the present article, we will review the effect of four kinds of phytoestrogens, genistein, resveratrol, kaempferol and 3,3′-diindolylmethane (DIM), on cancer progression via epithelial-mesenchymal transition (EMT). As shown in Figure 1, genistein, kaempferol and resveratrol are phenolic compounds: genistein and kaempferol are isoflavones, having a common flavone structure; resveratrol is a derivative of diphenylethane; and DIM is an active indole compound originated from indole-3-carbinol (I3C), an inactive form of indole. They are actively-studied phytoestrogens that have great potential to display anti-cancer effects.

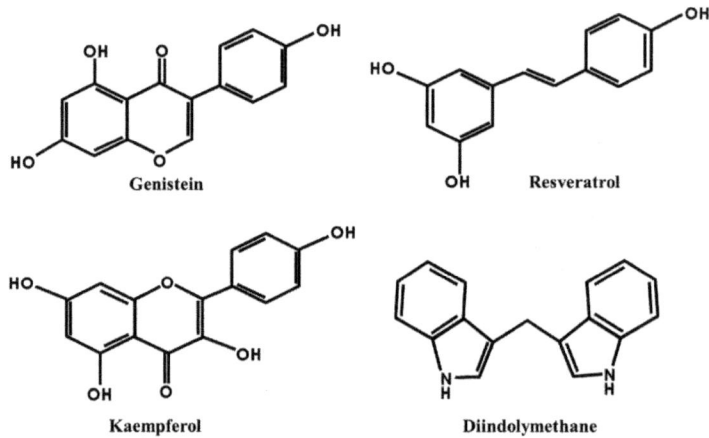

Figure 1. Chemical structures of phytoestrogens, genistein, resveratrol, kaempferol and 3,3′-diindolylmethane (DIM).

EMT plays a key role in tumor progression. The cancer cells undergoing EMT increase the expression levels of cell motility-related proteins and show enhanced migration and invasion to other sites of the body, resulting in cancer metastasis [33]. It has been found that crucial EMT markers, such as E-cadherin and Snail, are identified to secure positive evidence of EMT, and several signaling pathways are associated with the induction of EMT process in cancer cells, including Not-1 and TGF-beta signalings. In the next section, the importance of EMT in cancer progression, diverse EMT markers and related signaling pathways are briefly introduced to further highlight the impact of these phytoestrogens in chemoprevention against cancer.

2. Epithelial-Mesenchymal Transition in Cancer Metastasis

According to the World Cancer Report 2014 of the World Health Organization (WHO), about 14.1 million new cases of cancer occurred globally in 2012, leading to 14.6% of all human deaths. Approximately 90% of all cancer-related deaths are reported to be associated with tumor metastasis [34]. The chance of having an invasive cancer in one's lifetime is estimated to be 42% for men and 38% for women [35]. The characteristic of cancer malignancy and metastasis is the propagation of primary tumors through migrating to and invading the surrounding tissues [36]. Tumor cells have the potential to invade other tissues and to form metastasis through multiple steps known as malignant progression [37].

The program responsible for profound modification for metastasis that enables detaching from the junctions and dismissing the lateral cell-cell adhesions of cancer cells is EMT [38]. Tumor cells undergoing EMT display unique phenotypes, express higher levels of cell motility proteins and show promoted migration and invasion abilities [39].

EMT is associated with several major characteristics of cellular phenotypes. Through this process, epithelial cells change the morphology from a cobblestone-like monolayer with apical basal polarity to flat and spindle-shaped mesenchymal cells in the absence of polarization to gain the ability to move [40].

To acquire the moving ability, epithelial cells lose their ability to maintain the entire junction complex that connects them to the neighboring ones, of which basolateral surfaces are regularly spaced through membrane-associated specialized junctions [41]. In the process in which epithelial cells are switched to mesenchymal cells, the formation of a space where a barrier and rigidity are maintained is inhibited due to the lack of intercellular junctions [42]. A number of cells undergoing EMT develop interactions with the extracellular environment in localized areas of the carcinoma, where they involve the loss of intercellular cohesion, the disruption of extracellular matrix (ECM), modifications of the cytoskeleton, increased motility and invasion into the extracellular space [43]. Particularly, EMT is related to the expression of extracellular matrix proteases, such as matrix metalloproteases (MMPs) and urokinase-type plasminogen activator (uPA), which can degrade the ECM linked to the plasma membrane and localized to invadopodia during metastasis [44–46]. Meanwhile, EMT is a reversible process that can convert to its inverse process, called the mesenchymal-epithelial transition (MET). The cells undergoing MET increase cell-cell adhesion and return to epithelial phenotypes, which also play a role during embryonic development and pathological processes [47,48]. As a result, primary cancer cells lose cell-cell adhesion via EMT by *E*-cadherin repression, break through the basement membrane and enter the bloodstream through intravasation. Later, the circulating tumor cells exit the bloodstream to migrate to the specific metastatic sites where they undergo MET for clonal outgrowth [49].

A diversity of molecules associated with the process of EMT has been established, and some crucial molecules have been employed as biological markers to determine the process. As a typical molecule in the adherent junctions, *E*-cadherin, which is a transmembrane glycoprotein of the type I cadherin family and a crucial epithelial marker, has been found to inactivate and repress tumor progression by maintaining intact cell-cell interactions and inhibiting cell mobility, invasion and metastasis in human cancer [50,51]. The loss of *E*-cadherin expression is allowed for a critical step in the progression of invasive carcinoma by causing the EMT event. The loss of many epithelial markers (including *E*-cadherin, occludin, claudins and beta-catenin) induces the expression of mesenchymal markers (including *N*-cadherin, Snail, vimentin, R-cadherin and cadherin-11) and acquisition of mesenchymal characteristics, such as cell motility and invasion [52,53]. Additionally, the zinc-finger transcription factors, including Snail, Slug, and ZEB 1 and 2, have been shown to induce the EMT process by directly binding to the E-box of the *E*-cadherin promoter and suppressing the activity of *E*-cadherin [54], while Twist, another type of *E*-cadherin repressor, indirectly downregulates *E*-cadherin transcription [55].

It has been found that crucial EMT markers are associated with several signaling pathways in the induction of the EMT process in cancer cells. Snail has been found to be a critical factor in TGF-β signaling to resist cell death and to inhibit apoptosis [56]. As the most important factor that triggers EMT, TGF-β1 mediates the EMT process via numerous intracellular signaling pathways, including the Smad pathway, mitogen-activated protein kinases (MAPK), PI3K/Akt and small GTPases in HL-60 leukemia, Panc-1 human pancreatic and MDA-MB-231 breast cancer cells [57–59]. The overactivation of the TGF-β pathway in hepatocellular carcinoma cells confers a mesenchymal-like and an increased migratory capacity to the cells and finally contributes to tumor progression thorough the crosstalk with the chemokine CXCL12 pathway in liver tumor cells [60]. The Wnt signaling pathway activates β-catenin and several EMT-inducing transcription factors, such as Slug and Twist [34]. This signaling pathway induces cancer cell proliferation, motility and intravasation. Furthermore, in *in vivo* studies, the Wnt pathway displayed an important role in regulating EMT progression of colorectal cancer [61], and the Wnt-β-catenin was activated in the mesenchyme of the cardiac cushion during EMT in zebrafish and mouse embryo [62,63]. The cancer development of organs has been regulated by Notch-1 signaling, which directly promotes Snail, Slug and NF-κB in BxPC-3 human pancreatic cancer cell [64]. Notch-1 signaling also induced cell proliferation, survival and EMT by increasing NF-κB transcriptional activity in many human malignancies, including pancreatic and breast cancer cells [65,66]. In addition, the Hedgehog (Hh) signaling pathway is currently considered as a therapeutic target for anti-cancer treatment, because this pathway is abnormally activated in various types of cancer and contributes to tumor metastasis by inducing EMT. The misregulation of Hh signaling has been implicated as an important mediator in human pancreatic carcinoma, and specifically the sonic hedgehog pathway promotes metastasis and lymphangiogenesis via activation of Akt, EMT and the MMP-9 pathway in gastric cancers [67,68].

Recently, microRNAs (miRNAs) are being considered as an important regulator of EMT in various cancer cells. They incompletely bind to the 3′untranslated region (3′UTR) of mRNA to inhibit the translations [69]. The incomplete accordance between miRNAs and their targets allows the chances for miRNAs to control multiple genes. Moreover, miRNAs have been shown to play a crucial role during caner development and progression via the modulation of the expression of their target mRNA transcripts [70]. High miR-34a levels stimulate MET by reversing Snail and TCF-β-induced EMT [71]. As a negative regulator in the EMT process, miR-125a induced MET by the epidermal growth factor receptor (EGFR) signaling pathway [72]. miR-506 suppresses EMT, cell proliferation, migration and invasion by upregulating *E*-cadherin [73]. miR-138 also has a role in the inhibition of EMT and invasion in SKOV-3 ovarian cancer cells [74]. Another miRNA, miR-30a, was reported to suppress cell motility and EMT via targeting the expression of mesenchymal markers, thereby increasing the epithelial marker in A549 lung and BGC-823 gastric cancer cells [75,76]. On the other hand, miR-106a is associated with cell proliferation and tumor differentiation, and miR-7 is linked with metastasis and EMT [77]. In the case of miR-10b, its high expression upregulated the EMT and the expression of EMT-related proteins in metastatic tumors and induced the changed spindle-like morphology, cell migration and overexpression of *N*-cadherin, Snail, Slug and Twist [78,79]. Specifically, the miR-200 family is significant for reducing the ZEB levels, cell migration and TGF-β-induced EMT [80,81]. Expression of the miR-200 family was increased or decreased in the process of metastasis: miR-200 was downregulated in the EMT process, while it was upregulated during the re-epithelialization of distal metastasis [82].

3. Phytoestrogens and Their Actions on Cancer Cells Undergoing EMT

Since the dysregulation of proteins in signaling pathways involved in EMT is associated with cancer progression, they could be potentially targeted as prognostic markers or therapeutic targets of cancer metastasis [83]. Phytoestrogens having a chemopreventive effect on cancer progression seem to inhibit the EMT process through various channels.

3.1. Genistein

Genistein (40,5,7-trihydroxyisoflavone), having a heterocyclic diphenolic structure similar to estrogen, is a typical isoflavonoid found in a number of plants, including soybeans, peas, lentils and other beans [84,85]. As a phytoestrogen, it has an ability to bind and activate ERs, preferentially ERβ rather than ERα [86,87]. The higher binding affinity for ERβ of genistein has been associated with its action as an estrogen antagonist and having chemopreventive activity in estrogen-responsive cancers [88].

Anti-proliferative and chemopreventive effects of genistein have been extensively investigated in hormone-related, as well as non-hormone-related cancers in which genistein affects many crucial cellular functions related to carcinogenesis, including cell proliferation, apoptosis, cell cycle progression, migration, metastasis and invasion [89]. Recent studies have elucidated that genistein may have the potential to inhibit cancer metastasis by specifically regulating the EMT process via diverse signaling pathways.

Notch-1 signaling is an important pathway to upregulate the expression of EMT markers, ZEB1 and 2, Slug and vimentin, leading to the EMT, migration and drug resistance of pancreatic cancer cells [90,91]. In AsPC-1 pancreatic cancer cells, Notch-1 overexpression affected the expression of miRNAs: overexpression of Notch-1 led to increased expression of miR-21 and decreased expression of miR-200b. Re-expression of miR-200b led to inhibition of the EMT process by inducing decreased expression of ZEB1 and vimentin and increased expression of *E*-cadherin. Genistein treatment was found to attenuate the acquisition of EMT by AsPC-1 cells by promoting re-expression of miR-200b, which was repressed by Notch-1 signaling [90].

In a recent study, genistein suppressed the EMT of BG-1 ovarian cancer cells, which was activated by E2 and endocrine-disrupting chemicals, such as bisphenol A (BPA) and nonylphenol (NP), by downregulating the TGF-β pathway [92]. As a result, genistein not only suppressed the migration of BG-1 cells, but also diminished the expression of mesenchymal markers (vimentin) and metastasis markers (MMP-2 and cathepsin D) [92]. Genistein also effectively inhibited TGF-β-induced invasion and metastasis in the Panc-1 pancreatic cancer cell line through Smad4-dependent and independent pathways through p38 MAPK [58]. The EMT process was also reversed by genistein in the HepG2 hepatocellular carcinoma cell line by downregulating the nuclear factor of activated T cells 1 (NFAT1) [93]. NFAT1 was known to function in cell-autonomous actions, like invasion, migration, differentiation and proliferation in tumors [94]. Although the underlying mechanism has not been found yet, genistein suppressed invasive growth of LNCaP prostate cancer cells through the reversal of EMT, even at low concentrations (less than 15 micromoles/L genistein), which did not affect cell proliferation [95].

3.2. Resveratrol

Resveratrol (trans-3,4,5-trihydroxystilbene) is one of the stilbene phytoalexins, first found in the roots of the oriental medicinal plant *Polygonum cuspidatum* (Kojo-kon in Japanese) [96], and also exists in diverse vegetables, including berries, peanuts and red grape [97]. Resveratrol is known to be produced naturally when the plant is injured under attack by pathogens, such as bacteria or fungi [98]. Therefore, the proper infection of *Botrytis cinerea* (the fungus responsible for grey mold) is needed to obtain maximal concentrations of resveratrol within wine [99]. The characteristic of resveratrol as a phytoestrogen has been verified by its capability to mainly bind to ER and to regulate the transcription of estrogen-responsive target genes [100]. Many studies showed that resveratrol binds to ERβ with a higher affinity than to ERα, though it binds with 7000-fold lower affinity than E2, and that it acts as an agonist or an antagonist in the cells expressing ER [101,102]. Resveratrol can also regulate androgen receptor (AR)-mediated actions as a chemopreventive agent against prostate cancer: it inhibited androgen-stimulated cell growth and gene expression by repressing the expression and function of the AR in LNCaP prostate cancer cells [103].

In addition to its sex hormone-related actions, resveratrol has been found to be very helpful in inhibiting diabetes, heart disease and diverse cancers, because it possesses various bioactive properties, such as anti-oxidation, anti-proliferation, anti-inflammation and induction of apoptosis [104]. Specifically, the anti-oxidative efficacy of resveratrol to prevent the ROS generation and oxidative stress that may drive epithelial cells into an EMT program can be an effective characteristic of resveratrol to prevent the EMT of cancer cells. Actually, modulation of oxidative stress may be an efficient therapeutic tool for the inhibition of cancer progression [105]. Resveratrol inhibited the hypoxia-enhanced proliferation, invasion and EMT process in Saos-2 osteosarcoma cells via downregulation of the HIF-1α protein [106]. A previous study also revealed that resveratrol effectively suppressed the hypoxia-driven ROS-induced invasive and migratory ability of pancreatic cancer cells by inhibiting the Hh signaling pathway, which is able to regulate the EMT [107].

In addition, a recent finding indicated that resveratrol suppressed invasion and metastasis in gastric cancer by inhibiting the Hh signaling pathway and EMT [108]. In PC-3 and LNCaP prostate cancer cell lines, lipopolysaccharide (LPS) was used to trigger EMT, but resveratrol inhibited LPS-induced morphological changes, cell motility and invasiveness, the expression of EMT markers and inhibited the expression of glioma-associated oncogene homolog 1 (Gli1), suggesting that resveratrol has in part the ability to inhibit the EMT process through the Hh signaling pathway [109].

Similar to genistein, resveratrol abrogates the TGF-β1-induced EMT process for cancer progression. In LoVo colorectal cancer cells, resveratrol inhibited the invasive and migratory ability of LoVo cells, increased the expression of *E*-cadherin and repressed the expression of vimentin, via the inhibition of the TGF-β1/Smads signaling pathway [110]. Other studies also support the role of resveratrol in EMT inhibition. Xu *et al.* reported that resveratrol reversed EMT by inhibiting AKT signaling in pancreatic cancer [111]. Previously, AKT1 was also shown to promote EMT, as well as to increase metastasis in squamous cancer and sarcoma [112]. On the contrary, AKT1, but not AKT2 and AKT3, inhibited EMT in breast cancer, depressing Twist1 activation [112]. In the MCF-7 breast cancer cell line, resveratrol was found to inhibit EGF-induced EMT via inhibition of the EGF-mediated Erk pathway activation [113]. The role of resveratrol in inhibiting EMT induction was demonstrated in A549 lung cancer cells [114].

3.3. Kaempferol

Kaempferol (3,5,7-trihydroxy-2-(4-hydroxyphenyl)-4*H*-1-benzopyran-4-one) is one of the flavonoids found in many edible plants, like tea, broccoli, cabbage, beans and tomato, and its name was derived due to its specific source of the rhizome of *Kaempferi galangal* L., known as a popular traditional aromatic plant [115,116]. As one of the phytoestrogens due to its polyphenolic structure, kaempferol also exerts anti-proliferative and anti-carcinogenic actions though ER, AR, the aryl hydrocarbon receptor (AhR) and the progesterone receptor (PR) signaling pathways in many types of cancer [117–119].

Apoptosis is one of the main pathways for kaempferol to induce the anti-carcinogenic effect. In some cells, kaempferol induced apoptosis by stimulating the enzyme activity of caspases, which are a group of the cysteine proteases that are important initiators or effectors of the apoptosis process [65]. For caspase-3, it activates apoptosis by inducing DNA fragmentation and chromatin condensation in nucleus [120]. Kaempferol decreased the mitochondria potential by the stimulation of caspase-3 activity, resulting in the apoptosis of human lung non-small carcinoma cells [121,122]. In caspase independent pathways, kaempferol also promoted apoptosis by translocating apoptosis-inducing factor (AIF) into nucleus. AIF, which exists mainly in the space between the inner and outer mitochondrial membrane, was translocated into nuclei by kaempferol to induce nuclear condensation and large-scale DNA fragmentation [123,124].

Kaempferol also seems to inhibit cancer invasion and metastasis via the inhibition of EMT. Specifically for lung cancer, kaempferol was well known to suppress cancer migration, invasiveness and metastasis by modulating the expression of EMT proteins. Kaempferol significantly reduced the expression of MMP and mesenchymal markers and repressed metastasis and EMT by the TGF-β-dependent signaling pathway in non-small cell lung cancer [125]. In A549 lung cancer cells, kaempferol exerted the suppression of TGF-β1-induced EMT, migration and metastasis by blocking Smad3 as an important mediator of TGF-β signaling. In this study, PI3K/Akt signaling stimulated EMT and cell migration by directly phosphorylating Smas3, but kaempferol repressed EMT and cell migration by inhibiting Akt1-mediated phosphorylation of Smad3 [126]. The effect of kaempferol on EMT in relation to cancer progression has not been fully demonstrated yet, except for lung cancer. Although not in cancer, kaempferol was found to alleviate fibrotic airway remodeling via bronchial EMT by modulating protease-activated receptor-1 (PAR1) activation, which was entailed by TGF-β, suggesting that it may be a potential therapeutic agent targeting asthmatic airway constriction [127]. Since kaempferol is considered to have obvious anti-EMT efficacies, it will be applicable to other cancers for the prevention of cancer metastasis induced by EMT. According to a recent review on ribosomal S6 kinase (RSK) isoforms, the synthetic version of kaempferol, kaempferol-glycoside, can effectively target the invasion and metastasis of cancer by inhibiting RKS isoforms that promote invasion and tumor metastasis [128].

3.4. DIM

Brassica vegetables, such as cabbage, Brussels sprouts and broccoli, are rich in indole-3-carbinol (I3C), which is one of the active phytonutrients and a precursor of many different compounds, especially DIM. I3C undergoes acid-catalyzed dehydration and polymerization to be DIM in an acidic environment [129].

DIM has been shown to abrogate the proliferation of human cancer cells of prostate, breast, colon, ovary and pancreas [130]. Especially, it has the potential role of AhR signaling by acting as one of the selective AhR modulators (SAhRMs) in mammary carcinogenesis prevention [131]. With respect to anti-cancer mechanisms, DIM encourages apoptotic cell death by downregulation of NF-κB, survivin and Bcl-2 as anti-apoptotic factors and upregulation of Bax, a pro-apoptotic factor [132,133]. One surprising ability of DIM as a potential anti-cancer compound is its selective induction of apoptosis in cancerous cells, but not in normal cells [134]. Furthermore, DIM activates cell cycle modulators, p21 and p27, leading to G1 cell cycle arrest in breast, ovarian, prostate and colon cancer cell lines [132,135].

In addition to the protective effects of DIM against tumorigenesis, DIM has more effects in inhibiting chemotaxis and metastasis by inactivating of CXCR4 and CXCL12 at low concentrations by affecting AhR and ER in carcinogenesis [136]. The urokinase plasminogen activator (uPA) system is confirmed to have potential effects in cell migration, angiogenesis, cancer invasion and metastasis. Some studies have shown that DIM can downregulate uPA in the inhibition of tumor progression in prostate and breast cancer [137,138]. Furthermore, DIM inhibited cell proliferation, migration and metastasis by directly inactivating vascular endothelial growth factor (VEGF) and MMP and by involving the degradation of the basement membrane in original vessels, endothelial cell activation and migration [139].

DIM has been reported to repress tumor malignancy via inhibiting the EMT process. DIM deterred EMT in prostate cancer cells by blocking AR signaling and the expression of prostate-specific antigen (PSA), an AR-target gene, resulting in the reduction of the expression of EMT markers, ZEB1, *N*-cadherin and fibronectin [140]. DIM upregulated the protein expression of *E*-cadherin and downregulated the protein expression of vimentin by attenuating miR-92a and the NF-κB receptor activator. The PI3K/Akt/mTOR/NF-κB signaling pathway has been known to induce migration, invasion and EMT. However, DIM reversed EMT by modulating the PI3K/Akt/mTOR/NF-κB signaling. DIM also decreased the expression of mesenchymal markers ZEB1, vimentin and Slug and changed the morphology of pancreatic cancer cells into the epithelial form [141,142].

Recently, miRNA has been shown to have a crucial role in the initiation, progression and metastasis of cancer, and DIM has been shown to function as miRNA regulators affecting cancer metastasis and growth by modulating EMT [143]. DIM can repair the miRNA deformity and inhibit the mutation of miRNA in vinyl carbamate-induced mouse lung cancer [144]. According to a recent study by Li *et al.*, DIM treatment decreased the expression of vimentin, ZEB1 and Slug as EMT markers with reversal of EMT in pancreatic cancer cells via the increased expression of miR-200 [145]. DIM promoted the expression of the let-7 family, while it reversed the tumor progression and EMT by preventing ZEB1 [145], and it conducted the reduction of enhancer of zeste homolog 2 (EZH2) intervening in EMT phenotypic cells [146]. In a recent study, DIM also showed depression of miR-34a and inhibition of cancer growth via AR signaling *in vivo* [147]. In addition, DIM upregulated the protein expression of *E*-cadherin and downregulated the protein expression of vimentin by attenuating miR-92a and the NF-κB receptor activator [148]. In this way, DIM as an miRNA controller can lead to the repression of EMT and to a new therapy in prostate cancer.

Li *et al.* also demonstrated that DIM could be effective against pancreatic cancer by reversing the EMT phenomenon by upregulating the expression of miR-200 and the let-7 family, which are typically lost in many other cancers [145]. Wu *et al.* showed that DIM inhibited the metastasis of nasopharyngeal carcinoma (NPC) and EMT by promoting the expression of mesenchymal markers, Snail and Slug [149]. Moreover, DIM reversed *E*-cadherin expression in NPC, which experienced the EMT process, and DIM-induced *E*-cadherin expression has confirmed as having a significant positive correlation with the long-time outcome of NPC patients [150].

4. Conclusions

The phytoestrogens have generated considerable interests as alternatives for hormone replacement therapy (HRT) and chemopreventive compounds for recent years, because they have diverse bioactivities, low toxicity for the human body and easy acceptance as dietary supplements.

In the present study, we suggested significant roles of phytoestrogens in the inhibition of cancer progression via EMT, which is a crucial process inducing cancer migration, invasion and metastasis. Genistein, resveratrol, kaempferol and DIM have been evaluated to effectively repress the EMT process in diverse cancers by affecting the signaling pathways and the expression of EMT-related markers, as shown in the diagram of Figure 2. Specific signaling pathways and EMT markers regulated by genistein, resveratrol, kaempferol and DIM, are summarized in Table 1. In addition, since these phytoestrogens have been also reported to exert anti-carcinogenic effects by controlling cell cycle, cell proliferation and apoptosis in hormone-related, as well as non-hormone-related cancers, they may be considered as effective anti-cancer agents to inhibit the whole process of cancer progression.

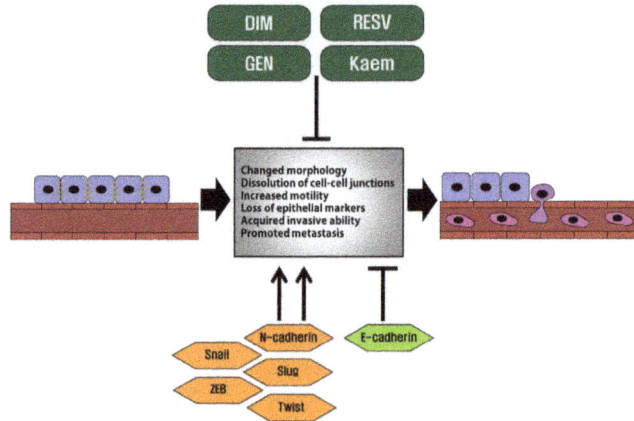

Figure 2. Schematic diagram of the EMT process and the roles of phytoestrogens, genistein, resveratrol, kaempferol and DIM, in the regulation of the EMT process in cancer metastasis. EMT plays a key role in tumor progression. The cells undergoing EMT show *E*-cadherin repression, but increased expression of EMT markers, such as Snail, Slug and vimentin, and cell motility-related proteins, including MMPs and uPA. As a result, they gain enhanced migration and invasion capabilities: primary cancer cells lose cell-cell adhesion, break through the basement membrane and enter the bloodstream through intravasation. Later, the circulating tumor cells exit the bloodstream to migrate to the specific metastatic sites, where they undergo MET for clonal outgrowth. On the other hand, genistein, resveratrol, kaempferol and DIM may inhibit cancer metastasis by repressing the EMT process through affecting the signaling pathways associated with EMT and regulating the expression of EMT markers.

Table 1. Potential signal transductions related to EMT targeted by dietary phytoestrogens.

Phytoestrogen	EMT-Related Signalings	Reference
Genistein	TGF-β, Smad, PI3K, Akt, NF-kB, Notch-1, MAPK, ER	[58,90,92]
Resveratrol	Hedgehog, TGF-β, Smad, AKT, EGF	[107,108,110,111,113,114]
Kaempferol	ER, AR, AhR, PR, TGF-β, Smad3, PI3K/Akt, RAF/ERK	[117–119,125,126,128]
Diindolylmethane	AR, PI3K/Akt/mTOR/NF-κB, Hedgehog, miR-200, RANKL, β-catenin	[140,141,143,145,147,148,150]

Nevertheless, controversies about the chemopreventive effect of phytoestrogens have existed because their cancer promoting effects have been also reported. Several phytoestrogens, including genistein and resveratrol, were known to have a biphasic effect, especially in hormone-dependent cancers: based on the specific concentration, cancer cell growth is stimulated at lower concentrations and inhibited at higher concentrations [19]. More skepticism regarding the chemopreventive effects of phytoestrogens came from the *in vitro* concentration of phytoestrogen being generally 10 μM, and plasma concentrations of >10 μM cannot be attainable by the dietary intake of phytoestrogen-containing food; thus, low levels of phytoestrogen, rather, may stimulate cancer growth. However, this insistence is still incompetent to abrogate the experimental and epidemiological evidence of phytoestrogens supporting anti-proliferative and anti-carcinogenic effects. Accordingly, more investigations of the relationship between the hazardous and chemopreventive effects of phytoestrogens in the body are needed at this stage. Besides, the poor bioavailability of phytoestrogens, such as resveratrol, in the body is another pitfall in the application of phytoestrogens, which has to be improved [97]. The toxicological issues posed by phytoestrogens also should be considered in their applications. For instance, the possibility that serum genistein concentrations found in soy-fed infants may induce thymic and immune abnormalities was raised [151]. High doses of genistein (>5 µM) may act as a topoisomerase II inhibitor and a DNA damaging genotoxin [152]. Genistein was reported to be capable of altering the toxicological behaviors of the endocrine-active pesticide methoxychlor and likely other endocrine-active compounds, as well [153]. In this way, the adverse effects of phytoestrogens on normal cellular function have been found. As with many other compounds, there are many pros and cons associated with phytoestrogens, and thus, it is urgently needed to shape the development of guidelines for the use of phytoestrogens for maximizing health benefits, as well as minimizing the adverse effects [154].

From this review, we may suggest that phytoestrogens are potent compounds that abrogate the cell migration, invasion and metastasis of cancer by effectively suppressing the EMT process. Phytoestrogen treatments are a promising way to prevent cancer development as safer alternatives for the natural strategies against cancers, though we need more research and interest for going forward to clinical trials.

Acknowledgments: This work was supported by a National Research Foundation of Korea (NRF) grant funded by the Ministry of Education, Science and Technology (MEST) of the Republic of Korea (2014R1A1A2055295). This work was supported by a grant from the Next-Generation BioGreen 21 Program (No. PJ011355-2015), Rural Development Administration, Republic of Korea.

Conflicts of Interest: None of the authors have any conflicts of interest to declare.

References

1. Cragg, G.M.; Grothaus, P.G.; Newman, D.J. Impact of natural products on developing new anti-cancer agents. *Chem. Rev.* **2009**, *109*, 3012–3043. [CrossRef] [PubMed]
2. Fitter, A. Ecology. Making allelopathy respectable. *Science* **2003**, *301*, 1337–1338. [CrossRef] [PubMed]
3. Acharya, A.; Das, I.; Singh, S.; Saha, T. Chemopreventive properties of indole-3-carbinol, diindolylmethane and other constituents of cardamom against carcinogenesis. *Recent Pat. Food Nutr. Agric.* **2010**, *2*, 166–177. [CrossRef] [PubMed]
4. Sporn, M.B.; Suh, N. Chemoprevention of cancer. *Carcinogenesis* **2000**, *21*, 525–530. [CrossRef] [PubMed]
5. Cragg, G.M.; Newman, D.J. Plants as a source of anti-cancer agents. *J. Ethnopharmacol.* **2005**, *100*, 72–79. [CrossRef] [PubMed]
6. Thakur, V.S.; Deb, G.; Babcook, M.A.; Gupta, S. Plant phytochemicals as epigenetic modulators: Role in cancer chemoprevention. *AAPS J.* **2014**, *16*, 151–163. [CrossRef] [PubMed]
7. Allred, K.F.; Yackley, K.M.; Vanamala, J.; Allred, C.D. Trigonelline is a novel phytoestrogen in coffee beans. *J. Nutr.* **2009**, *139*, 1833–1838. [CrossRef] [PubMed]
8. Santen, R.J.; Boyd, N.F.; Chlebowski, R.T.; Cummings, S.; Cuzick, J.; Dowsett, M.; Easton, D.; Forbes, J.F.; Key, T.; Hankinson, S.E.; *et al.* Critical assessment of new risk factors for breast cancer: Considerations for development of an improved risk prediction model. *Endocr. Relat. Cancer* **2007**, *14*, 169–187. [CrossRef] [PubMed]
9. Giacalone, P.L.; Daures, J.P.; Ouafik, L.; Martin, P.M.; Laffargue, F.; Maudelonde, T. Steroids and adrenomedullin growth patterns in human ovarian cancer cells: Estrogenic-regulation assay. *Gynecol. Oncol.* **2003**, *91*, 651–656. [CrossRef]
10. Chung, S.H.; Franceschi, S.; Lambert, P.F. Estrogen and ERalpha: Culprits in cervical cancer? *Trends Endocrinol. Metab.* **2010**, *21*, 504–511. [CrossRef] [PubMed]
11. Hwang, K.A.; Kang, N.H.; Yi, B.R.; Lee, H.R.; Park, M.A.; Choi, K.C. Genistein, a soy phytoestrogen, prevents the growth of BG-1 ovarian cancer cells induced by 17beta-estradiol or bisphenol A via the inhibition of cell cycle progression. *Int. J. Oncol.* **2013**, *42*, 733–740. [PubMed]

12. Ahn, H.N.; Jeong, S.Y.; Bae, G.U.; Chang, M.; Zhang, D.; Liu, X.; Pei, Y.; Chin, Y.W.; Lee, J.; Oh, S.R.; *et al.* Selective estrogen receptor modulation by larrea nitida on MCF-7 Cell proliferation and immature rat uterus. *Biomol. Ther.* **2014**, *22*, 347–354. [CrossRef] [PubMed]
13. Shang, Y.; Brown, M. Molecular determinants for the tissue specificity of SERMs. *Science* **2002**, *295*, 2465–2468. [CrossRef] [PubMed]
14. Kuiper, G.G.; Lemmen, J.G.; Carlsson, B.; Corton, J.C.; Safe, S.H.; van der Saag, P.T.; van der Burg, B.; Gustafsson, J.A. Interaction of estrogenic chemicals and phytoestrogens with estrogen receptor beta. *Endocrinology* **1998**, *139*, 4252–4263. [PubMed]
15. Zhao, E.; Mu, Q. Phytoestrogen biological actions on Mammalian reproductive system and cancer growth. *Sci. Pharm.* **2011**, *79*, 1–20. [CrossRef] [PubMed]
16. Harris, D.M.; Besselink, E.; Henning, S.M.; Go, V.L.; Heber, D. Phytoestrogens induce differential estrogen receptor alpha- or Beta-mediated responses in transfected breast cancer cells. *Exp. Biol. Med.* **2005**, *230*, 558–568.
17. Hsieh, C.Y.; Santell, R.C.; Haslam, S.Z.; Helferich, W.G. Estrogenic effects of genistein on the growth of estrogen receptor-positive human breast cancer (MCF-7) cells *in vitro* and *in vivo*. *Cancer Res.* **1998**, *58*, 3833–3838. [PubMed]
18. Ju, Y.H.; Fultz, J.; Allred, K.F.; Doerge, D.R.; Helferich, W.G. Effects of dietary daidzein and its metabolite, equol, at physiological concentrations on the growth of estrogen-dependent human breast cancer (MCF-7) tumors implanted in ovariectomized athymic mice. *Carcinogenesis* **2006**, *27*, 856–863. [CrossRef] [PubMed]
19. Magee, P.J.; Rowland, I.R. Phyto-oestrogens, their mechanism of action: Current evidence for a role in breast and prostate cancer. *Br. J. Nutr.* **2004**, *91*, 513–531. [CrossRef] [PubMed]
20. Rice, S.; Whitehead, S.A. Phytoestrogens and breast cancer—Promoters or protectors? *Endocr. Relat. Cancer* **2006**, *13*, 995–1015. [CrossRef] [PubMed]
21. Kirk, C.J.; Harris, R.M.; Wood, D.M.; Waring, R.H.; Hughes, P.J. Do dietary phytoestrogens influence susceptibility to hormone-dependent cancer by disrupting the metabolism of endogenous oestrogens? *Biochem. Soc. Trans.* **2001**, *29*, 209–216. [CrossRef] [PubMed]
22. Labrie, F.; Luu-The, V.; Lin, S.X.; Labrie, C.; Simard, J.; Breton, R.; Belanger, A. The key role of 17 β-hydroxysteroid dehydrogenases in sex steroid biology. *Steroids* **1997**, *62*, 148–158. [CrossRef]
23. Murkies, A.L.; Wilcox, G.; Davis, S.R. Clinical review 92, Phytoestrogens. *J. Clin. Endocrinol. Metab.* **1998**, *83*, 297–303. [PubMed]
24. Pilsakova, L.; Riecansky, I.; Jagla, F. The physiological actions of isoflavone phytoestrogens. *Physiol. Res.* **2010**, *59*, 651–664. [PubMed]
25. Hutchins, A.M.; Lampe, J.W.; Martini, M.C.; Campbell, D.R.; Slavin, J.L. Vegetables, fruits, and legumes: Effect on urinary isoflavonoid phytoestrogen and lignan excretion. *J. Am. Diet. Assoc.* **1995**, *95*, 769–774. [CrossRef]
26. Milder, I.E.; Arts, I.C.; van de Putte, B.; Venema, D.P.; Hollman, P.C. Lignan contents of Dutch plant foods: A database including lariciresinol, pinoresinol, secoisolariciresinol and matairesinol. *Br. J. Nutr.* **2005**, *93*, 393–402. [CrossRef] [PubMed]
27. Zamora-Ros, R.; Knaze, V.; Lujan-Barroso, L.; Kuhnle, G.G.; Mulligan, A.A.; Touillaud, M.; Slimani, N.; Romieu, I.; Powell, N.; Tumino, R.; *et al.* Dietary intakes and food sources of phytoestrogens in the European Prospective Investigation into Cancer and Nutrition (EPIC) 24-h dietary recall cohort. *Eur. J. Clin. Nutr.* **2012**, *66*, 932–941. [CrossRef] [PubMed]
28. Heinonen, S.; Nurmi, T.; Liukkonen, K.; Poutanen, K.; Wahala, K.; Deyama, T.; Nishibe, S.; Adlercreutz, H. *In vitro* metabolism of plant lignans: New precursors of mammalian lignans enterolactone and enterodiol. *J. Agric. Food Chem.* **2001**, *49*, 3178–3186. [CrossRef] [PubMed]
29. Boker, L.K.; Van der Schouw, Y.T.; De Kleijn, M.J.; Jacques, P.F.; Grobbee, D.E.; Peeters, P.H. Intake of dietary phytoestrogens by Dutch women. *J. Nutr.* **2002**, *132*, 1319–1328. [PubMed]
30. Manjanatha, M.; Shelton, S.; Bishop, M.; Lyn-Cook, L.; Aidoo, A. Dietary effects of soy isoflavones daidzein and genistein on 7,12-dimethylbenz[a]anthracene-induced mammary mutagenesis and carcinogenesis in ovariectomized Big Blue transgenic rats. *Carcinogenesis* **2006**, *27*, 1970–1979. [CrossRef] [PubMed]
31. Sobolev, V.S.; Horn, B.W.; Potter, T.L.; Deyrup, S.T.; Gloer, J.B. Production of stilbenoids and phenolic acids by the peanut plant at early stages of growth. *J. Agric. Food Chem.* **2006**, *54*, 3505–3511. [CrossRef] [PubMed]
32. Jang, M.; Cai, L.; Udeani, G.O.; Slowing, K.V.; Thomas, C.F.; Beecher, C.W.; Fong, H.H.; Farnsworth, N.R.; Kinghorn, A.D.; Mehta, R.G.; *et al.* Cancer chemopreventive activity of resveratrol, a natural product derived from grapes. *Science* **1997**, *275*, 218–220. [CrossRef] [PubMed]

33. Heerboth, S.; Housman, G.; Leary, M.; Longacre, M.; Byler, S.; Lapinska, K.; Willbanks, A.; Sarkar, S. EMT and tumor metastasis. *Clin. Transl. Med.* **2015**, *4*. [CrossRef] [PubMed]

34. Spano, D.; Heck, C.; De Antonellis, P.; Christofori, G.; Zollo, M. Molecular networks that regulate cancer metastasis. *Semin. Cancer Biol.* **2012**, *22*, 234–249. [CrossRef] [PubMed]

35. Siegel, R.L.; Miller, K.D.; Jemal, A. Cancer statistics, 2016. *CA Cancer J. Clin.* **2016**, *66*, 7–30. [CrossRef] [PubMed]

36. Voulgari, A.; Pintzas, A. Epithelial-mesenchymal transition in cancer metastasis: Mechanisms, markers and strategies to overcome drug resistance in the clinic. *Biochim. Biophys. Acta* **2009**, *1796*, 75–90. [CrossRef] [PubMed]

37. Hanahan, D.; Weinberg, R.A. The hallmarks of cancer. *Cell* **2000**, *100*, 57–70. [CrossRef]

38. Yap, A.S.; Brieher, W.M.; Gumbiner, B.M. Molecular and functional analysis of cadherin-based adherens junctions. *Annu. Rev. Cell Dev. Biol.* **1997**, *13*, 119–146. [CrossRef] [PubMed]

39. Onder, T.T.; Gupta, P.B.; Mani, S.A.; Yang, J.; Lander, E.S.; Weinberg, R.A. Loss of *E*-cadherin promotes metastasis via multiple downstream transcriptional pathways. *Cancer Res.* **2008**, *68*, 3645–3654. [CrossRef] [PubMed]

40. Boyer, B.; Thiery, J.P. Epithelium-mesenchyme interconversion as example of epithelial plasticity. *APMIS* **1993**, *101*, 257–268. [CrossRef] [PubMed]

41. Thiery, J.P. Cell adhesion in development: A complex signaling network. *Curr. Opin. Genet. Dev.* **2003**, *13*, 365–371. [CrossRef]

42. Miyoshi, J.; Takai, Y. Structural and functional associations of apical junctions with cytoskeleton. *Biochim. Biophys. Acta* **2008**, *1778*, 670–691. [CrossRef] [PubMed]

43. Yang, J.; Weinberg, R.A. Epithelial-mesenchymal transition: At the crossroads of development and tumor metastasis. *Dev. Cell* **2008**, *14*, 818–829. [CrossRef] [PubMed]

44. Friedl, P.; Wolf, K. Tumour-cell invasion and migration: Diversity and escape mechanisms. *Nat. Rev. Cancer* **2003**, *3*, 362–374. [CrossRef] [PubMed]

45. Polette, M.; Nawrocki-Raby, B.; Gilles, C.; Clavel, C.; Birembaut, P. Tumour invasion and matrix metalloproteinases. *Crit. Rev. Oncol. Hematol.* **2004**, *49*, 179–186. [CrossRef] [PubMed]

46. Son, H.; Moon, A. Epithelial-mesenchymal Transition and Cell Invasion. *Toxicol. Res.* **2010**, *26*, 245–252. [CrossRef] [PubMed]

47. Davies, J.A. Mesenchyme to epithelium transition during development of the mammalian kidney tubule. *Acta Anat.* **1996**, *156*, 187–201. [CrossRef] [PubMed]

48. Foroni, C.; Broggini, M.; Generali, D.; Damia, G. Epithelial-mesenchymal transition and breast cancer: Role, molecular mechanisms and clinical impact. *Cancer Treat. Rev.* **2012**, *38*, 689–697. [CrossRef] [PubMed]

49. Chaffer, C.L.; Weinberg, R.A. A perspective on cancer cell metastasis. *Science* **2011**, *331*, 1559–1564. [CrossRef] [PubMed]

50. Perl, A.K.; Wilgenbus, P.; Dahl, U.; Semb, H.; Christofori, G. A causal role for *E*-cadherin in the transition from adenoma to carcinoma. *Nature* **1998**, *392*, 190–193. [CrossRef] [PubMed]

51. Thiery, J.P.; Sleeman, J.P. Complex networks orchestrate epithelial-mesenchymal transitions. *Nat. Rev. Mol. Cell Biol.* **2006**, *7*, 131–142. [CrossRef] [PubMed]

52. Bailey, J.M.; Singh, P.K.; Hollingsworth, M.A. Cancer metastasis facilitated by developmental pathways: Sonic hedgehog, Notch, and bone morphogenic proteins. *J. Cell. Biochem.* **2007**, *102*, 829–839. [CrossRef] [PubMed]

53. Zeisberg, M.; Neilson, E.G. Biomarkers for epithelial-mesenchymal transitions. *J. Clin. Investig.* **2009**, *119*, 1429–1437. [CrossRef] [PubMed]

54. Batlle, E.; Sancho, E.; Franci, C.; Dominguez, D.; Monfar, M.; Baulida, J.; Garcia De Herreros, A. The transcription factor snail is a repressor of *E*-cadherin gene expression in epithelial tumour cells. *Nat. Cell Biol.* **2000**, *2*, 84–89. [CrossRef] [PubMed]

55. Yang, M.H.; Wu, M.Z.; Chiou, S.H.; Chen, P.M.; Chang, S.Y.; Liu, C.J.; Teng, S.C.; Wu, K.J. Direct regulation of TWIST by HIF-1alpha promotes metastasis. *Nat. Cell Biol.* **2008**, *10*, 295–305. [CrossRef] [PubMed]

56. Thiery, J.P. Epithelial-mesenchymal transitions in development and pathologies. *Curr. Opin. Cell Biol.* **2003**, *15*, 740–746. [CrossRef] [PubMed]

57. Bani-Hani, A.H.; Campbell, M.T.; Meldrum, D.R.; Meldrum, K.K. Cytokines in epithelial-mesenchymal transition: A new insight into obstructive nephropathy. *J. Urol.* **2008**, *180*, 461–468. [CrossRef] [PubMed]

58. Han, L.; Zhang, H.W.; Zhou, W.P.; Chen, G.M.; Guo, K.J. The effects of genistein on transforming growth factor-beta1-induced invasion and metastasis in human pancreatic cancer cell line Panc-1 *in vitro*. *Chin. Med. J.* **2012**, *125*, 2032–2040. [PubMed]

59. Zhang, Y.E. Non-Smad pathways in TGF-β signaling. *Cell Res.* **2009**, *19*, 128–139. [CrossRef] [PubMed]

60. Bertran, E.; Crosas-Molist, E.; Sancho, P.; Caja, L.; Lopez-Luque, J.; Navarro, E.; Egea, G.; Lastra, R.; Serrano, T.; Ramos, E.; *et al.* Overactivation of the TGF-β pathway confers a mesenchymal-like phenotype and CXCR4-dependent migratory properties to liver tumor cells. *Hepatology* **2013**, *58*, 2032–2044. [CrossRef] [PubMed]

61. Hu, T.H.; Yao, Y.; Yu, S.; Han, L.L.; Wang, W.J.; Guo, H.; Tian, T.; Ruan, Z.P.; Kang, X.M.; Wang, J.; *et al.* SDF-1/CXCR4 promotes epithelial-mesenchymal transition and progression of colorectal cancer by activation of the Wnt/beta-catenin signaling pathway. *Cancer Lett.* **2014**, *354*, 417–426. [CrossRef] [PubMed]

62. Gitler, A.D.; Lu, M.M.; Jiang, Y.Q.; Epstein, J.A.; Gruber, P.J. Molecular markers of cardiac endocardial cushion development. *Dev. Dyn.* **2003**, *228*, 643–650. [CrossRef] [PubMed]

63. Liebner, S.; Cattelino, A.; Gallini, R.; Rudini, N.; Iurlaro, M.; Piccolo, S.; Dejana, E. β-catenin is required for endothelial-mesenchymal transformation during heart cushion development in the mouse. *J. Cell Biol.* **2004**, *166*, 359–367. [CrossRef] [PubMed]

64. Wang, Z.; Banerjee, S.; Li, Y.; Rahman, K.M.; Zhang, Y.; Sarkar, F.H. Down-regulation of notch-1 inhibits invasion by inactivation of nuclear factor-kappaB, vascular endothelial growth factor, and matrix metalloproteinase-9 in pancreatic cancer cells. *Cancer Res.* **2006**, *66*, 2778–2784. [CrossRef] [PubMed]

65. Miyamoto, Y.; Maitra, A.; Ghosh, B.; Zechner, U.; Argani, P.; Iacobuzio-Donahue, C.A.; Sriuranpong, V.; Iso, T.; Meszoely, I.M.; Wolfe, M.S.; *et al.* Notch mediates TGF alpha-induced changes in epithelial differentiation during pancreatic tumorigenesis. *Cancer Cell* **2003**, *3*, 565–576. [CrossRef]

66. Shao, S.; Zhao, X.; Zhang, X.; Luo, M.; Zuo, X.; Huang, S.; Wang, Y.; Gu, S.; Zhao, X. Notch1 signaling regulates the epithelial-mesenchymal transition and invasion of breast cancer in a Slug-dependent manner. *Mol. Cancer* **2015**, *14*, 28. [CrossRef] [PubMed]

67. Thayer, S.P.; di Magliano, M.P.; Heiser, P.W.; Nielsen, C.M.; Roberts, D.J.; Lauwers, G.Y.; Qi, Y.P.; Gysin, S.; Fernandez-del Castillo, C.; Yajnik, V.; *et al.* Hedgehog is an early and late mediator of pancreatic cancer tumorigenesis. *Nature* **2003**, *425*, 851–856. [CrossRef] [PubMed]

68. Yoo, Y.A.; Kang, M.H.; Lee, H.J.; Kim, B.H.; Park, J.K.; Kim, H.K.; Kim, J.S.; Oh, S.C. Sonic hedgehog pathway promotes metastasis and lymphangiogenesis via activation of Akt, EMT, and MMP-9 pathway in gastric cancer. *Cancer Res.* **2011**, *71*, 7061–7070. [CrossRef] [PubMed]

69. Gregory, R.I.; Chendrimada, T.P.; Cooch, N.; Shiekhattar, R. Human RISC couples microRNA biogenesis and posttranscriptional gene silencing. *Cell* **2005**, *123*, 631–640. [CrossRef] [PubMed]

70. Filipowicz, W.; Jaskiewicz, L.; Kolb, F.A.; Pillai, R.S. Post-transcriptional gene silencing by siRNAs and miRNAs. *Curr. Opin. Struct. Biol.* **2005**, *15*, 331–341. [CrossRef] [PubMed]

71. Siemens, H.; Jackstadt, R.; Hunten, S.; Kaller, M.; Menssen, A.; Gotz, U.; Hermeking, H. miR-34 and SNAIL form a double-negative feedback loop to regulate epithelial-mesenchymal transitions. *Cell Cycle* **2011**, *10*, 4256–4271. [CrossRef] [PubMed]

72. Dahl, K.D.C.; Dahl, R.; Kruichak, J.N.; Hudson, L.G. The epidermal growth factor receptor responsive miR-125a represses mesenchymal morphology in ovarian cancer cells. *Neoplasia* **2009**, *11*, 1208–1215. [CrossRef]

73. Yang, D.; Sun, Y.; Hu, L.; Zheng, H.; Ji, P.; Pecot, C.V.; Zhao, Y.; Reynolds, S.; Cheng, H.; Rupaimoole, R.; *et al.* Integrated analyses identify a master microRNA regulatory network for the mesenchymal subtype in serous ovarian cancer. *Cancer Cell* **2013**, *23*, 186–199. [CrossRef] [PubMed]

74. Yeh, Y.M.; Chuang, C.M.; Chao, K.C.; Wang, L.H. MicroRNA-138 suppresses ovarian cancer cell invasion and metastasis by targeting SOX4 and HIF-1alpha. *Int. J. Cancer* **2013**, *133*, 867–878. [CrossRef] [PubMed]

75. Cheng, C.W.; Wang, H.W.; Chang, C.W.; Chu, H.W.; Chen, C.Y.; Yu, J.C.; Chao, J.I.; Liu, H.F.; Ding, S.L.; Shen, C.Y. MicroRNA-30a inhibits cell migration and invasion by downregulating vimentin expression and is a potential prognostic marker in breast cancer. *Breast Cancer Res. Treat.* **2012**, *134*, 1081–1093. [CrossRef] [PubMed]

76. Kumarswamy, R.; Mudduluru, G.; Ceppi, P.; Muppala, S.; Kozlowski, M.; Niklinski, J.; Papotti, M.; Allgayer, H. MicroRNA-30a inhibits epithelial-to-mesenchymal transition by targeting Snail1 and is downregulated in non-small cell lung cancer. *Int. J. Cancer* **2012**, *130*, 2044–2053. [CrossRef] [PubMed]

77. Liu, Z.; Gersbach, E.; Zhang, X.; Xu, X.; Dong, R.; Lee, P.; Liu, J.; Kong, B.; Shao, C.; Wei, J.J. miR-106a represses the Rb tumor suppressor p130 to regulate cellular proliferation and differentiation in high-grade serous ovarian carcinoma. *Mol. Cancer Res.* **2013**, *11*, 1314–1325. [CrossRef] [PubMed]

78. Ma, L.; Teruya-Feldstein, J.; Weinberg, R.A. Tumour invasion and metastasis initiated by microRNA-10b in breast cancer. *Nature* **2007**, *449*, 682–688. [CrossRef] [PubMed]

79. Zhang, L.; Sun, J.; Wang, B.; Ren, J.C.; Su, W.; Zhang, T. MicroRNA-10b triggers the Epithelial-Mesenchymal Transition (EMT) of laryngeal carcinoma Hep-2 cells by directly targeting the *E*-cadherin. *Appl. Biochem. Biotechnol.* **2015**, *176*, 33–44. [CrossRef] [PubMed]

80. Gregory, P.A.; Bert, A.G.; Paterson, E.L.; Barry, S.C.; Tsykin, A.; Farshid, G.; Vadas, M.A.; Khew-Goodall, Y.; Goodall, G.J. The miR-200 family and miR-205 regulate epithelial to mesenchymal transition by targeting ZEB1 and SIP1. *Nat. Cell Biol.* **2008**, *10*, 593–601. [CrossRef] [PubMed]

81. Korpal, M.; Lee, E.S.; Hu, G.; Kang, Y. The miR-200 family inhibits epithelial-mesenchymal transition and cancer cell migration by direct targeting of *E*-cadherin transcriptional repressors ZEB1 and ZEB2. *J. Biol. Chem.* **2008**, *283*, 14910–14914. [CrossRef] [PubMed]

82. Dykxhoorn, D.M. MicroRNAs and metastasis: Little RNAs go a long way. *Cancer Res.* **2010**, *70*, 6401–6406. [CrossRef] [PubMed]

83. Guarino, M. Epithelial-mesenchymal transition and tumour invasion. *Int. J. Biochem. Cell Biol.* **2007**, *39*, 2153–2160. [CrossRef] [PubMed]

84. Kim, S.H.; Kim, C.W.; Jeon, S.Y.; Go, R.E.; Hwang, K.A.; Choi, K.C. Chemopreventive and chemotherapeutic effects of genistein, a soy isoflavone, upon cancer development and progression in preclinical animal models. *Lab. Anim. Res.* **2014**, *30*, 143–150. [CrossRef] [PubMed]

85. Perabo, F.G.; Von Low, E.C.; Ellinger, J.; von Rucker, A.; Muller, S.C.; Bastian, P.J. Soy isoflavone genistein in prevention and treatment of prostate cancer. *Prostate Cancer Prostatic Dis.* **2008**, *11*, 6–12. [CrossRef] [PubMed]

86. Huang, J.; Nasr, M.; Kim, Y.; Matthews, H.R. Genistein inhibits protein histidine kinase. *J. Biol. Chem.* **1992**, *267*, 15511–15515. [PubMed]

87. Ju, Y.H.; Allred, K.F.; Allred, C.D.; Helferich, W.G. Genistein stimulates growth of human breast cancer cells in a novel, postmenopausal animal model, with low plasma estradiol concentrations. *Carcinogenesis* **2006**, *27*, 1292–1299. [CrossRef] [PubMed]

88. Pelekanou, V.; Leclercq, G. Recent insights into the effect of natural and environmental estrogens on mammary development and carcinogenesis. *Int. J. Dev. Biol.* **2011**, *55*, 869–878. [CrossRef] [PubMed]

89. Pavese, J.M.; Farmer, R.L.; Bergan, R.C. Inhibition of cancer cell invasion and metastasis by genistein. *Cancer Metastasis Rev.* **2010**, *29*, 465–482. [CrossRef] [PubMed]

90. Bao, B.; Wang, Z.; Ali, S.; Kong, D.; Li, Y.; Ahmad, A.; Banerjee, S.; Azmi, A.S.; Miele, L.; Sarkar, F.H. Notch-1 induces epithelial-mesenchymal transition consistent with cancer stem cell phenotype in pancreatic cancer cells. *Cancer Lett.* **2011**, *307*, 26–36. [CrossRef] [PubMed]

91. Kim, A.Y.; Kwak, J.H.; Je, N.K.; Lee, Y.H.; Jung, Y.S. Epithelial-mesenchymal transition is associated with acquired resistance to 5-Fluorocuracil in HT-29 colon cancer cells. *Toxicol. Res.* **2015**, *31*, 151–156. [CrossRef] [PubMed]

92. Kim, Y.S.; Choi, K.C.; Hwang, K.A. Genistein suppressed epithelial-mesenchymal transition and migration efficacies of BG-1 ovarian cancer cells activated by estrogenic chemicals via estrogen receptor pathway and downregulation of TGF-beta signaling pathway. *Phytomedicine* **2015**, *22*, 993–999. [CrossRef] [PubMed]

93. Dai, W.; Wang, F.; He, L.; Lin, C.; Wu, S.; Chen, P.; Zhang, Y.; Shen, M.; Wu, D.; Wang, C.; *et al.* Genistein inhibits hepatocellular carcinoma cell migration by reversing the epithelial-mesenchymal transition: Partial mediation by the transcription factor NFAT1. *Mol. Carcinog.* **2015**, *54*, 301–311. [CrossRef] [PubMed]

94. Liu, J.F.; Zhao, S.H.; Wu, S.S. Depleting NFAT1 expression inhibits the ability of invasion and migration of human lung cancer cells. *Cancer Cell Int.* **2013**, *13*, 41. [CrossRef] [PubMed]

95. Zhang, L.L.; Li, L.; Wu, D.P.; Fan, J.H.; Li, X.; Wu, K.J.; Wang, X.Y.; He, D.L. A novel anti-cancer effect of genistein: Reversal of epithelial mesenchymal transition in prostate cancer cells. *Acta Pharmacol. Sin.* **2008**, *29*, 1060–1068. [CrossRef] [PubMed]

96. Nonomura, S.; Kanagawa, H.; Makimoto, A. Chemical Constituents of Polygonaceous Plants. I. Studies on the Components of Ko-J O-Kon. (Polygonum Cuspidatum Sieb. Et Zucc.). *J. Pharm. Soc. Jpn.* **1963**, *83*, 988–990.

97. Carter, L.G.; D'Orazio, J.A.; Pearson, K.J. Resveratrol and cancer: Focus on *in vivo* evidence. *Endocr. Relat. Cancer* **2014**, *21*, R209–R225. [CrossRef] [PubMed]

98. Fremont, L. Biological effects of resveratrol. *Life Sci.* **2000**, *66*, 663–673. [CrossRef]
99. Adrian, M.; Rajaei, H.; Jeandet, P.; Veneau, J.; Bessis, R. Resveratrol oxidation in botrytis cinerea conidia. *Phytopathology* **1998**, *88*, 472–476. [CrossRef] [PubMed]
100. Ashby, J.; Tinwell, H.; Pennie, W.; Brooks, A.N.; Lefevre, P.A.; Beresford, N.; Sumpter, J.P. Partial and weak oestrogenicity of the red wine constituent resveratrol: Consideration of its superagonist activity in MCF-7 cells and its suggested cardiovascular protective effects. *J. Appl. Toxicol.* **1999**, *19*, 39–45. [CrossRef]
101. Bowers, J.L.; Tyulmenkov, V.V.; Jernigan, S.C.; Klinge, C.M. Resveratrol acts as a mixed agonist/antagonist for estrogen receptors alpha and beta. *Endocrinology* **2000**, *141*, 3657–3667. [PubMed]
102. Le Corre, L.; Chalabi, N.; Delort, L.; Bignon, Y.J.; Bernard-Gallon, D.J. Resveratrol and breast cancer chemoprevention: Molecular mechanisms. *Mol. Nutr. Food Res.* **2005**, *49*, 462–471. [CrossRef] [PubMed]
103. Mitchell, S.H.; Zhu, W.; Young, C.Y. Resveratrol inhibits the expression and function of the androgen receptor in LNCaP prostate cancer cells. *Cancer Res.* **1999**, *59*, 5892–5895. [PubMed]
104. Pervaiz, S. Resveratrol: From grapevines to mammalian biology. *FASEB J.* **2003**, *17*, 1975–1985. [CrossRef] [PubMed]
105. Leone, L.; Mazzetta, F.; Martinelli, D.; Valente, S.; Alimandi, M.; Raffa, S.; Santino, I. Klebsiella pneumoniae Is Able to Trigger Epithelial-Mesenchymal Transition Process in Cultured Airway Epithelial Cells. *PLoS ONE* **2016**, *11*, e0146365. [CrossRef] [PubMed]
106. Sun, Y.; Wang, H.; Liu, M.; Lin, F.; Hua, J. Resveratrol abrogates the effects of hypoxia on cell proliferation, invasion and EMT in osteosarcoma cells through downregulation of the HIF-1α protein. *Mol. Med. Rep.* **2015**, *11*, 1975–1981. [CrossRef] [PubMed]
107. Li, W.; Cao, L.; Chen, X.; Lei, J.; Ma, Q. Resveratrol inhibits hypoxia-driven ROS-induced invasive and migratory ability of pancreatic cancer cells via suppression of the Hedgehog signaling pathway. *Oncol. Rep.* **2016**, *35*, 1718–1726. [CrossRef] [PubMed]
108. Gao, Q.; Yuan, Y.; Gan, H.Z.; Peng, Q. Resveratrol inhibits the hedgehog signaling pathway and epithelial-mesenchymal transition and suppresses gastric cancer invasion and metastasis. *Oncol. Lett.* **2015**, *9*, 2381–2387. [CrossRef] [PubMed]
109. Li, J.; Chong, T.; Wang, Z.; Chen, H.; Li, H.; Cao, J.; Zhang, P.; Li, H. A novel anticancer effect of resveratrol: Reversal of epithelialmesenchymal transition in prostate cancer cells. *Mol. Med. Rep.* **2014**, *10*, 1717–1724. [PubMed]
110. Ji, Q.; Liu, X.; Han, Z.; Zhou, L.; Sui, H.; Yan, L.; Jiang, H.; Ren, J.; Cai, J.; Li, Q. Resveratrol suppresses epithelial-to-mesenchymal transition in colorectal cancer through TGF-β1/Smads signaling pathway mediated Snail/E-cadherin expression. *BMC Cancer* **2015**, *15*, 97. [CrossRef] [PubMed]
111. Xu, Q.; Zong, L.; Chen, X.; Jiang, Z.; Nan, L.; Li, J.; Duan, W.; Lei, J.; Zhang, L.; Ma, J.; *et al.* Resveratrol in the treatment of pancreatic cancer. *Ann. N. Y. Acad. Sci.* **2015**, *1348*, 10–19. [CrossRef] [PubMed]
112. Li, C.W.; Xia, W.; Lim, S.O.; Hsu, J.L.; Huo, L.; Wu, Y.; Li, L.Y.; Lai, C.C.; Chang, S.S.; Hsu, Y.H.; *et al.* AKT1 inhibits epithelial-to-mesenchymal transition in breast cancer through phosphorylation-dependent Twist1 degradation. *Cancer Res.* **2016**. [CrossRef] [PubMed]
113. Vergara, D.; Valente, C.M.; Tinelli, A.; Siciliano, C.; Lorusso, V.; Acierno, R.; Giovinazzo, G.; Santino, A.; Storelli, C.; Maffia, M. Resveratrol inhibits the epidermal growth factor-induced epithelial mesenchymal transition in MCF-7 cells. *Cancer Lett.* **2011**, *310*, 1–8. [CrossRef] [PubMed]
114. Wang, H.; Zhang, H.; Tang, L.; Chen, H.; Wu, C.; Zhao, M.; Yang, Y.; Chen, X.; Liu, G. Resveratrol inhibits TGF-β1-induced epithelial-to-mesenchymal transition and suppresses lung cancer invasion and metastasis. *Toxicology* **2013**, *303*, 139–146. [CrossRef] [PubMed]
115. Calderon-Montano, J.M.; Burgos-Moron, E.; Perez-Guerrero, C.; Lopez-Lazaro, M. A review on the dietary flavonoid kaempferol. *Mini Rev. Med. Chem.* **2011**, *11*, 298–344. [CrossRef] [PubMed]
116. Huang, L.; Yagura, T.; Chen, S. Sedative activity of hexane extract of *Keampferia galanga* L. and its active compounds. *J. Ethnopharmacol.* **2008**, *120*, 123–125. [CrossRef] [PubMed]
117. Boam, T. Anti-androgenic effects of flavonols in prostate cancer. *Ecancermedicalscience* **2015**, *9*, 585. [CrossRef] [PubMed]
118. Puppala, D.; Gairola, C.G.; Swanson, H.I. Identification of kaempferol as an inhibitor of cigarette smoke-induced activation of the aryl hydrocarbon receptor and cell transformation. *Carcinogenesis* **2007**, *28*, 639–647. [CrossRef] [PubMed]

119. Toh, M.F.; Mendonca, E.; Eddie, S.L.; Endsley, M.P.; Lantvit, D.D.; Petukhov, P.A.; Burdette, J.E. Kaempferol exhibits progestogenic effects in ovariectomized rats. *J. Steroids Horm. Sci.* **2014**, *5*, 136. [PubMed]

120. Kang, J.W.; Kim, J.H.; Song, K.; Kim, S.H.; Yoon, J.H.; Kim, K.S. Kaempferol and quercetin, components of Ginkgo biloba extract (EGb 761), induce caspase-3-dependent apoptosis in oral cavity cancer cells. *Phytother. Res.* **2010**, *24* (Suppl. S1), S77–S82. [CrossRef] [PubMed]

121. Niering, P.; Michels, G.; Watjen, W.; Ohler, S.; Steffan, B.; Chovolou, Y.; Kampkotter, A.; Proksch, P.; Kahl, R. Protective and detrimental effects of kaempferol in rat H4IIE cells: Implication of oxidative stress and apoptosis. *Toxicol. Appl. Pharmacol.* **2005**, *209*, 114–122. [CrossRef] [PubMed]

122. Samhan-Arias, A.K.; Martin-Romero, F.J.; Gutierrez-Merino, C. Kaempferol blocks oxidative stress in cerebellar granule cells and reveals a key role for reactive oxygen species production at the plasma membrane in the commitment to apoptosis. *Free Radic. Biol. Med.* **2004**, *37*, 48–61. [CrossRef] [PubMed]

123. Leung, H.W.; Lin, C.J.; Hour, M.J.; Yang, W.H.; Wang, M.Y.; Lee, H.Z. Kaempferol induces apoptosis in human lung non-small carcinoma cells accompanied by an induction of antioxidant enzymes. *Food Chem. Toxicol.* **2007**, *45*, 2005–2013. [CrossRef] [PubMed]

124. Liou, A.K.; Clark, R.S.; Henshall, D.C.; Yin, X.M.; Chen, J. To die or not to die for neurons in ischemia, traumatic brain injury and epilepsy: A review on the stress-activated signaling pathways and apoptotic pathways. *Prog. Neurobiol.* **2003**, *69*, 103–142. [CrossRef]

125. Liang, S.Q.; Marti, T.M.; Dorn, P.; Froment, L.; Hall, S.R.; Berezowska, S.; Kocher, G.; Schmid, R.A.; Peng, R.W. Blocking the epithelial-to-mesenchymal transition pathway abrogates resistance to anti-folate chemotherapy in lung cancer. *Cell Death Dis.* **2015**, *6*, e1824. [CrossRef] [PubMed]

126. Jo, E.; Park, S.J.; Choi, Y.S.; Jeon, W.K.; Kim, B.C. Kaempferol suppresses transforming growth factor-β1-Induced epithelial-to-mesenchymal transition and migration of A549 lung cancer cells by inhibiting Akt1-mediated phosphorylation of Smad3 at Threonine-179. *Neoplasia* **2015**, *17*, 525–537. [CrossRef] [PubMed]

127. Gong, J.H.; Cho, I.H.; Shin, D.; Han, S.Y.; Park, S.H.; Kang, Y.H. Inhibition of airway epithelial-to-mesenchymal transition and fibrosis by kaempferol in endotoxin-induced epithelial cells and ovalbumin-sensitized mice. *Lab. Investig.* **2014**, *94*, 297–308. [CrossRef] [PubMed]

128. Sulzmaier, F.J.; Ramos, J.W. RSK isoforms in cancer cell invasion and metastasis. *Cancer Res.* **2013**, *73*, 6099–6105. [CrossRef] [PubMed]

129. Anderton, M.J.; Manson, M.M.; Verschoyle, R.D.; Gescher, A.; Lamb, J.H.; Farmer, P.B.; Steward, W.P.; Williams, M.L. Pharmacokinetics and tissue disposition of indole-3-carbinol and its acid condensation products after oral administration to mice. *Clin. Cancer Res.* **2004**, *10*, 5233–5241. [CrossRef] [PubMed]

130. Banerjee, S.; Kong, D.; Wang, Z.; Bao, B.; Hillman, G.G.; Sarkar, F.H. Attenuation of multi-targeted proliferation-linked signaling by 3,3'-diindolylmethane (DIM): From bench to clinic. *Mutat. Res.* **2011**, *728*, 47–66. [CrossRef] [PubMed]

131. Marques, M.; Laflamme, L.; Benassou, I.; Cissokho, C.; Guillemette, B.; Gaudreau, L. Low levels of 3,3'-diindolylmethane activate estrogen receptor alpha and induce proliferation of breast cancer cells in the absence of estradiol. *BMC Cancer* **2014**, *14*, 524. [CrossRef] [PubMed]

132. Gong, Y.; Sohn, H.; Xue, L.; Firestone, G.L.; Bjeldanes, L.F. 3,3'-Diindolylmethane is a novel mitochondrial H(+)-ATP synthase inhibitor that can induce p21(Cip1/Waf1) expression by induction of oxidative stress in human breast cancer cells. *Cancer Res.* **2006**, *66*, 4880–4887. [CrossRef] [PubMed]

133. Hong, C.; Firestone, G.L.; Bjeldanes, L.F. Bcl-2 family-mediated apoptotic effects of 3,3'-diindolylmethane (DIM) in human breast cancer cells. *Biochem. Pharmacol.* **2002**, *63*, 1085–1097. [CrossRef]

134. Li, Y.; Chinni, S.R.; Sarkar, F.H. Selective growth regulatory and pro-apoptotic effects of DIM is mediated by AKT and NF-kappaB pathways in prostate cancer cells. *Front. Biosci.* **2005**, *10*, 236–243. [CrossRef] [PubMed]

135. Choi, H.J.; Lim do, Y.; Park, J.H. Induction of G1 and G2/M cell cycle arrests by the dietary compound 3,3'-diindolylmethane in HT-29 human colon cancer cells. *BMC Gastroenterol.* **2009**, *9*, 39. [CrossRef] [PubMed]

136. Hsu, E.L.; Chen, N.; Westbrook, A.; Wang, F.; Zhang, R.; Taylor, R.T.; Hankinson, O. CXCR4 and CXCL12 down-regulation: A novel mechanism for the chemoprotection of 3,3'-diindolylmethane for breast and ovarian cancers. *Cancer Lett.* **2008**, *265*, 113–123. [CrossRef] [PubMed]

137. Ahmad, A.; Kong, D.; Sarkar, S.H.; Wang, Z.; Banerjee, S.; Sarkar, F.H. Inactivation of uPA and its receptor uPAR by 3,3'-diindolylmethane (DIM) leads to the inhibition of prostate cancer cell growth and migration. *J. Cell. Biochem.* **2009**, *107*, 516–527. [CrossRef] [PubMed]

138. Ahmad, A.; Kong, D.; Wang, Z.; Sarkar, S.H.; Banerjee, S.; Sarkar, F.H. Down-regulation of uPA and uPAR by 3,3'-diindolylmethane contributes to the inhibition of cell growth and migration of breast cancer cells. *J. Cell. Biochem.* **2009**, *108*, 916–925. [CrossRef] [PubMed]

139. Kong, D.; Li, Y.; Wang, Z.; Banerjee, S.; Sarkar, F.H. Inhibition of angiogenesis and invasion by 3,3'-diindolylmethane is mediated by the nuclear factor-kappaB downstream target genes MMP-9 and uPA that regulated bioavailability of vascular endothelial growth factor in prostate cancer. *Cancer Res.* **2007**, *67*, 3310–3319. [CrossRef] [PubMed]

140. Kong, D.; Sethi, S.; Li, Y.; Chen, W.; Sakr, W.A.; Heath, E.; Sarkar, F.H. Androgen receptor splice variants contribute to prostate cancer aggressiveness through induction of EMT and expression of stem cell marker genes. *Prostate* **2015**, *75*, 161–174. [CrossRef] [PubMed]

141. Ahmad, A.; Biersack, B.; Li, Y.; Kong, D.; Bao, B.; Schobert, R.; Padhye, S.B.; Sarkar, F.H. Targeted regulation of PI3K/Akt/mTOR/NF-kappaB signaling by indole compounds and their derivatives: Mechanistic details and biological implications for cancer therapy. *Anticancer Agents Med. Chem.* **2013**, *13*, 1002–1013. [CrossRef] [PubMed]

142. Lamouille, S.; Derynck, R. Emergence of the phosphoinositide 3-kinase-Akt-mammalian target of rapamycin axis in transforming growth factor-beta-induced epithelial-mesenchymal transition. *Cells Tissues Organs* **2011**, *193*, 8–22. [CrossRef] [PubMed]

143. Sethi, S.; Li, Y.; Sarkar, F.H. Regulating miRNA by natural agents as a new strategy for cancer treatment. *Curr. Drug Targets* **2013**, *14*, 1167–1174. [CrossRef] [PubMed]

144. Melkamu, T.; Zhang, X.; Tan, J.; Zeng, Y.; Kassie, F. Alteration of microRNA expression in vinyl carbamate-induced mouse lung tumors and modulation by the chemopreventive agent indole-3-carbinol. *Carcinogenesis* **2010**, *31*, 252–258. [CrossRef] [PubMed]

145. Li, Y.; VandenBoom, T.G., 2nd; Kong, D.; Wang, Z.; Ali, S.; Philip, P.A.; Sarkar, F.H. Up-regulation of miR-200 and let-7 by natural agents leads to the reversal of epithelial-to-mesenchymal transition in gemcitabine-resistant pancreatic cancer cells. *Cancer Res.* **2009**, *69*, 6704–6712. [CrossRef] [PubMed]

146. Kong, D.; Heath, E.; Chen, W.; Cher, M.L.; Powell, I.; Heilbrun, L.; Li, Y.; Ali, S.; Sethi, S.; Hassan, O.; *et al.* Loss of let-7 up-regulates EZH2 in prostate cancer consistent with the acquisition of cancer stem cell signatures that are attenuated by BR-DIM. *PLoS ONE* **2012**, *7*, e33729. [CrossRef] [PubMed]

147. Watson, G.W.; Beaver, L.M.; Williams, D.E.; Dashwood, R.H.; Ho, E. Phytochemicals from cruciferous vegetables, epigenetics, and prostate cancer prevention. *AAPS J.* **2013**, *15*, 951–961. [CrossRef] [PubMed]

148. Li, Y.; Kong, D.; Ahmad, A.; Bao, B.; Sarkar, F.H. Targeting bone remodeling by isoflavone and 3,3'-diindolylmethane in the context of prostate cancer bone metastasis. *PLoS ONE* **2012**, *7*, e33011. [CrossRef] [PubMed]

149. Wu, T.; Chen, C.; Li, F.; Chen, Z.; Xu, Y.; Xiao, B.; Tao, Z. 3,3'-Diindolylmethane inhibits the invasion and metastasis of nasopharyngeal carcinoma cells and by regulation of epithelial mesenchymal transition. *Exp. Ther. Med.* **2014**, *7*, 1635–1638. [PubMed]

150. Xu, L.; Jiang, Y.; Zheng, J.; Xie, G.; Li, J.; Shi, L.; Fan, S. Aberrant expression of β-catenin and E-cadherin is correlated with poor prognosis of nasopharyngeal cancer. *Hum. Pathol.* **2013**, *44*, 1357–1364. [CrossRef] [PubMed]

151. Yellayi, S.; Naaz, A.; Szewczykowski, M.A.; Sato, T.; Woods, J.A.; Chang, J.; Segre, M.; Allred, C.D.; Helferich, W.G.; Cooke, P.S. The phytoestrogen genistein induces thymic and immune changes: A human health concern? *Proc. Natl. Acad. Sci. USA* **2002**, *99*, 7616–7621. [CrossRef] [PubMed]

152. Klein, C.B.; King, A.A. Genistein genotoxicity: Critical considerations of *in vitro* exposure dose. *Toxicol. Appl. Pharmacol.* **2007**, *224*, 1–11. [CrossRef] [PubMed]

153. You, L. Phytoestrogen genistein and its pharmacological interactions with synthetic endocrine-active compounds. *Curr. Pharm. Des.* **2004**, *10*, 2749–2757. [CrossRef] [PubMed]

154. Patisaul, H.B.; Jefferson, W. The pros and cons of phytoestrogens. *Front. Neuroendocrinol.* **2010**, *31*, 400–419. [CrossRef] [PubMed]

toxins

MDPI

Article

Pueraria mirifica Exerts Estrogenic Effects in the Mammary Gland and Uterus and Promotes Mammary Carcinogenesis in Donryu Rats

Anna Kakehashi [1,*], Midori Yoshida [2,†], Yoshiyuki Tago [1], Naomi Ishii [1], Takahiro Okuno [1], Min Gi [1] and Hideki Wanibuchi [1]

[1]   Department of Molecular Pathology, Osaka City University Graduate School of Medicine,
     1-4-3 Asahi-machi, Abeno-ku, Osaka 545-8585, Japan; Yoshiyuki_Tago@kn.kaneka.co.jp (Y.T.);
     naomi-u@med.osaka-cu.ac.jp (N.I.); m2026860@med.osaka-cu.ac.jp (T.O.); mwei@med.osaka-cu.ac.jp (M.G.);
     wani@med.osaka-cu.ac.jp (H.W.)
[2]   Division of Pathology, Biological Safety Research Center, National Institute of Health Sciences,
     Ministry of Health, Labour and Welfare, 1-18-1 Kamiyoga, Setagaya-ku, Tokyo 158-8501, Japan
*   Correspondence: anna@med.osaka-cu.ac.jp; Tel.: +81-6-6645-3737
†   Present address: Food Safety Commission, Cabinet Office, Government of Japan, 5-2-20 Akasaka, Minato-ku,
     Tokyo 107-6122, Japan; midori.yoshida@cao.go.jp.

Academic Editor: Carmela Fimognari
Received: 3 June 2016; Accepted: 13 September 2016; Published: 4 November 2016

Abstract: *Pueraria mirifica* (PM), a plant whose dried and powdered tuberous roots are now widely used in rejuvenating preparations to promote youthfulness in both men and women, may have major estrogenic influence. In this study, we investigated modifying effects of PM at various doses on mammary and endometrial carcinogenesis in female Donryu rats. Firstly, PM administered to ovariectomized animals at doses of 0.03%, 0.3%, and 3% in a phytoestrogen-low diet for 2 weeks caused significant increase in uterus weight. Secondly, a 4 week PM application to non-operated rats at a dose of 3% after 7,12-dimethylbenz[a]anthracene (DMBA) initiation resulted in significant elevation of cell proliferation in the mammary glands. In a third experiment, postpubertal administration of 0.3% (200 mg/kg body weight (b.w.)/day) PM to 5-week-old non-operated animals for 36 weeks following initiation of mammary and endometrial carcinogenesis with DMBA and N-ethyl-N'-nitro-N-nitrosoguanidine (ENNG), respectively, resulted in significant increase of mammary adenocarcinoma incidence. A significant increase of endometrial atypical hyperplasia multiplicity was also observed. Furthermore, PM at doses of 0.3%, and more pronouncedly, at 1% induced dilatation, hemorrhage and inflammation of the uterine wall. In conclusion, postpubertal long-term PM administration to Donryu rats exerts estrogenic effects in the mammary gland and uterus, and at a dose of 200 mg/kg b.w./day was found to promote mammary carcinogenesis initiated by DMBA.

Keywords: *Pueraria mirifica*; mammary gland; uterus; carcinogenesis; estrogenic activity; Donryu rat

1. Introduction

Pueraria mirifica (PM), also known as white Kwao Krua, is a plant found in northern and northeastern Thailand which belongs to the family of Leguminosae, and the soy, bean, and pea subfamily Papilionoideae. Previously, PM application for treatment of a range of conditions, including those related to the aging process, has been reported [1]. Dried and powdered tuberous roots of PM contain at least 17 chemical compounds with estrogenic biological activities, usually divided into three groups: the first group includes teniso flavonoids such as genistin, genistein, daidzein, daidzin, kwakhurin, kwakhurin hydrate, tuberosin, puerarin, mirificin, and puemiricarpene [2,3];

the second group of coumestans comprises coumestrol, mirificoumestan, mirificoumestan glycol, and mirificoumestan hydrate; and the third features chromenes, such as miroestrol, deoxymiroestrol, and isomiroestrol [2]. All these substances are phytoestrogens with structures similar to that of 17 β-estradiol. Miroestrol, the phytoestrogen with the highest estrogenic activity among all those isolated from PM, is considered similar to estriol, which is considered the safest estrogen for humans [4–7]. Furthermore, PM has been reported to contain phytoestrogens like β-sitosterol, stigmasterol, campesterol, as well as the cytotoxic non-phytoestrogen spinasterol [3,8]. Puerarin might account for about half of the total isoflavone content of PM, with lower amounts of genistin and daidzin present, all these being glycoside forms which can be partially hydrolyzed in the intestine by C-glycosyl bond cleavage to give the respective aglycoside forms: genistein, daidzein, and daidzein [2].

Nowadays, PM is available in tablets, extracts, creams, sprays, and powdered forms, so that it can be added to other medicinal preparations or herbs, and individual conditions require different applications and dosages [1]. It can be readily obtained from internet resources in many countries, including the USA and Japan, and is primarily used for supporting memory, smoothing the skin, increasing hair growth, improving appetite, and providing relief for ailments like osteoporosis and even cancer [9–12]. Continuous administration of PM at 20–100 mg/day for 6 months, or at 100–200 mg/day for 12 months, was found to help women having menopause symptoms, while no significant changes were detected in their hepatic, hematologic, and renal functions [3]. Pretreatment with PM at a high dose (1000 mg/kg body weight (b.w.)/day) for 4 weeks was shown to suppress the development of mammary tumors induced by 7,12-dimethylbenz[a]anthracene (DMBA) in Sprague–Dawley rats [13]. However, the effect of long-term administration at different doses has not yet been clarified in detail. Based on available studies, a safe dosage of PM as a dietary supplement for humans was suggested at 1–2 mg/kg b.w./day or about 50–100 mg/day [3]. Nowadays, doses of 20–100 mg/day are commonly used, but in some cases 200–900 mg/day or even higher (up to 3000 mg/day) are applied. Until now, no serious side effects have been recorded with the prescribed safe dosage, although at high doses PM may cause epilepsy, diabetes, asthma, and migraine [3].

Despite the data on the benefits of PM, there are reasons for concern that a herb which exhibits strong estrogen-like properties may stimulate the growth of existing estrogen-sensitive breast or endometrial tumors, pointing out questions such as: what is a safe dose? Previously, nanomolar concentrations of genistein, present in PM, was shown to induce acid ceramidase (ASAH1) transcription through a GPR30-dependent, pertussis toxin-sensitive pathway that requires the activation of c-Src and extracellular signal regulated kinase 1/2 (ERK1/2), thus stimulating breast cancer cell growth [14]. Recently, we further demonstrated that postpubertal administration of soy isoflavones at estrogenic doses promotes mammary and endometrial carcinogenesis in Donryu rats [15]. These data call into question the safety of long-term exposure to phytoestrogens with regard to effects on the mammary gland and endometrium. It is of particular importance that concentrations of PM which might exert promoting effects on mammary gland and uterine carcinogenesis be determined. Therefore, the present study was carried out to investigate the modifying effects of various doses of PM on mammary and uterine endometrial carcinogenesis using the Donryu rat model.

2. Results

2.1. Estrogenic Effect of Test Compounds (Short-Term Experiment 1)

After ovariectomy, PM treatment at doses of 0.3% and 3% and isoflavone aglycon (IA) treatment at a dose of 0.2% resulted in cornification, evident on examination of vaginal smears (Figure 1).

The 0.3% PM treatment was found to exert even stronger estrogenic activity than 0.2% IA. Mean relative uterus weights were significantly elevated in 0.03% ($0.13\% \pm 0.01\%$, $p < 0.05$), 0.3% ($0.31\% \pm 0.03\%$, $p < 0.05$), 3% PM ($0.35\% \pm 0.08\%$, $p < 0.05$) and 0.2% IA ($0.19\% \pm 0.04\%$, $p < 0.05$) administered ovariectomized rats as compared to the control ($0.08\% \pm 0.01\%$). Thus, weak, medium, and strong estrogenic activities of PM at doses of 0.03% (20 mg/kg b.w./day),

0.3% (200 mg/kg b.w./day), and 3% (2000 mg/kg b.w./day), respectively, were demonstrated in the rat uterus. In this experiment, significant decreases of body weights were observed in the 0.3% and 3% PM groups ($p < 0.001$) as well as 0.2% IA group ($p < 0.01$). Because of the very strong estrogenic effect of 3% PM detected in the short-term study, in the succeeding long-term experiment the highest dose was changed from 3% to 1%. Thus, the test doses of PM in experiment 3 were set as: low, 0.03%; medium, 0.3%; and high, 1%.

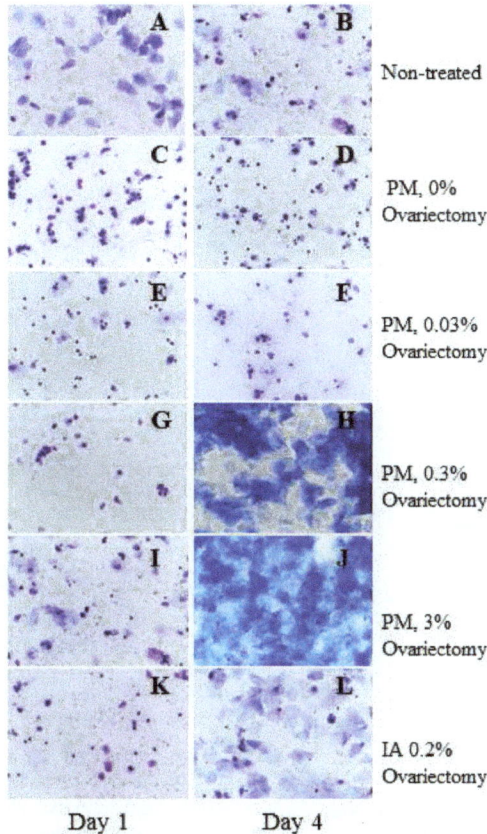

Figure 1. Vaginal cytology for non-treated (**A,B**) and ovariectomized (**C–L**) female Donryu rats administered *Pueraria mirifica* (PM) and isoflavone aglycon (IA) for the first 4 days. Animals were given PM at doses of 0 (**C,D**), 0.03 (**E,F**), 0.3 (**G,H**) and 3% (**I,J**), or 0.2% IA (**K,L**) 2 weeks after the ovariectomy. Vaginal smears were obtained daily before and after starting the treatment, dried and stained with an aqueous solution of methylene blue. In the ovariectomized rats the absence of cyclicity was confirmed by castration smears typical of diestrus. Note that 0.3% and 3% PM as well as 0.2% IA rats exerted estrogenic activities confirmed by cornification which was similar to estrus status.

2.2. Cell Proliferation in the Mammary Gland (Short-Term Experiment 2)

Bromodeoxyuridine (BrdU) immunohistochemistry revealed a dose-dependent increase of cell proliferation in the terminal end buds of mammary glands of PM-treated rats after 4 weeks of administration (Figure 2). Significant elevation of the number of BrdU positively (BrdU$^+$)-stained cells was noted in 3% PM ($p < 0.001$) and 0.2% IA groups ($p < 0.001$), as compared to the DMBA control group.

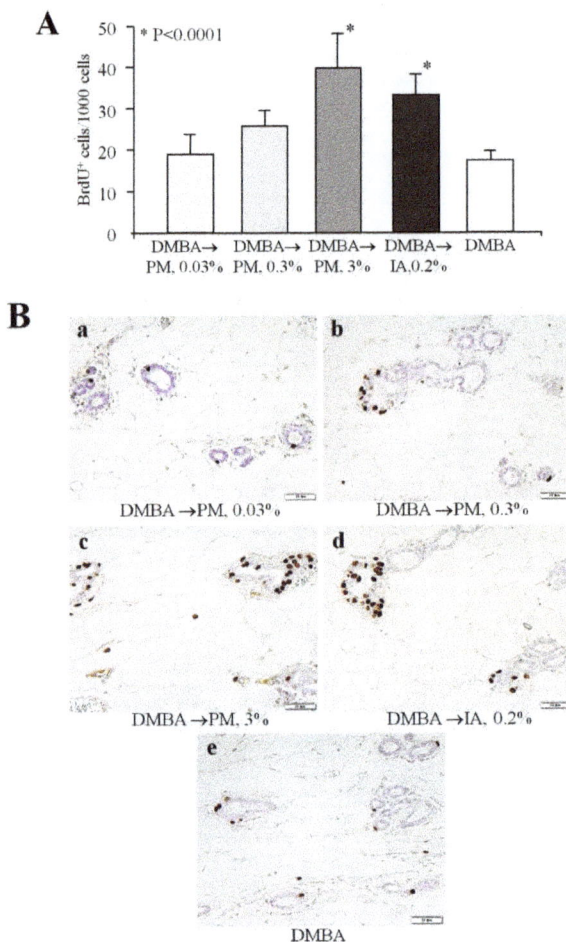

Figure 2. Bromodeoxyuridine (BrdU) labeling indices in the mammary glands of rats administered PM and IA after the 7,12-dimethylbenz[a]anthracene (DMBA) initiation (**A**); representative pictures of BrdU immunohistochemistry of mammary glands of rats (**B**). Note the dose-dependent induction of cell proliferation by the short-term application of PM as compared to the DMBA initiation control rats.

2.3. Long-Term Study (Experiment 3)

2.3.1. Body, Organ Weights, Food, and Water Consumption

Rat body weight curves are presented in Figure 3A. Body weights of 0.3% PM-, 1% PM-, and 0.2% IA-administered rats were lower than in the initiation control group, with significant differences detected at the termination of the experiment. No variation in food intake was observed among groups but decreased water consumption was noted in PM- and IA-treated animals.

Significant increases of relative uterus weight were found with 0.3 and 1% PM, and a trend for increase was found in the vehicle 1% PM group, as compared to the respective control groups (Supplementary Materials Table S1). Relative liver weights were significantly decreased in 0.3 and 1% PM-treated rats. Moreover, significant elevation of relative kidney weights was found in the 1% PM group. In addition, significant increase of relative thymus weights at 0.03% PM and elevation of relative adrenal weights in the vehicle 1% PM group as compared to the respective controls were noted.

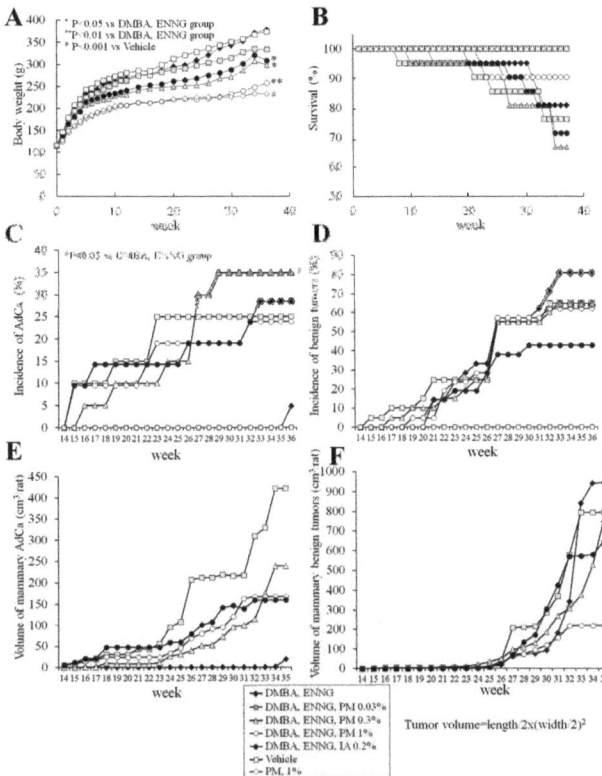

Figure 3. Body weight (**A**) and survival (**B**) curves of female Donryu rats in experiment 3; incidences of mammary adenocarcinomas (**C**), benign tumors (**D**), and volumes of mammary adenocarcinomas (**E**) and benign tumors (**F**). Note the significant decreases of body weights in the 0.3 and 1% PM- and 0.2% IA-treated Donryu rats. Trends for decrease in survival were found for 0.03 and 0.3% PM and 0.2% IA groups. Significant increases of mammary adenocarcinoma incidence were observed in 0.3% PM- and 0.2% IA-treated rats. Adenocarcinomas in PM- and IA-treated rats appeared earlier, and their volumes were higher than in the initiation control group. Development of benign tumors in the initiation control group starting at week 32 was evident. Mammary tumors were larger in animals with higher body weights.

2.3.2. Survival

Changes in rat survival are shown in Figure 3B. Four animals in the initiation control, 6 rats in the 0.03% PM, 7 rats in the 0.3% PM, 2 rats in the 1% PM, and 6 rats in the 0.2% IA groups died during the study. One rat each in the vehicle control, 0.03% PM, and 0.3% PM groups were found dead with no discernible cause. The apparent causes of death in the initiation control group were zymbal gland tumor (1 rat) and malignant lymphoma/leukemia (3 rats). In contrast, causes of death in the 0.03%, 0.3%, 1% PM and 0.2% IA groups were mammary adenocarcinomas (0.03% PM, 2 rats; 0.3% PM, 2 rats; 1% PM, 1 rat; IA, 3 rats), with bleeding from large necrotic mammary tumors (0.03% PM, 2 rats; 0.3% PM, 1 rat; 1% PM, 1 rat), lymphoma/leukemia (0.03% PM, 2 rats; 0.3% PM, 1 rat; IA, 2 rats), thymoma (0.3% PM, 1 rat), and uterine carcinoma (IA, 1 rat). Two rats in the 0.03% PM group, 1 rat receiving 0.3% PM, and 3 animals in the IA-treated group featured metastasis from mammary adenocarcinomas in the lung.

The first rat was found dead at week 9, from the 0.03% PM group, and the cause of death was lymphoma/leukemia. Subsequently, one rat in the 0.3% PM group at week 10 and one rat from the initiation control group at week 15 died from lymphoma/leukemia. The numbers of animals in the 0.03 and 0.3% PM- and IA-treated groups, but not the 1% PM group, then started to decrease mostly due to the development of mammary adenocarcinomas (Figure 3B). Survival rates of 0.03 and 0.3% PM- and IA-administered rats showed a nonsignificant trend for decrease at the termination of the experiment.

2.3.3. Histopathological Analysis of Mammary Glands

Data for changes in mammary gland adenocarcinoma and benign tumor incidences, multiplicities, and volumes—with representative pictures—are shown in Figure 3C–F, Figure 4A(c,d) and Table 1.

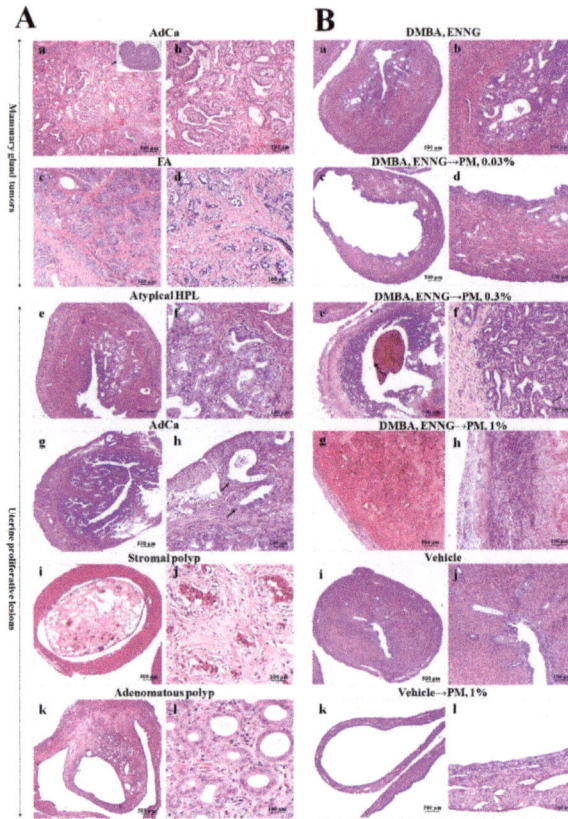

Figure 4. (**A**) Histopathological changes observed in mammary glands (**a–d**) and uteri (**e–l**) of initiation control, PM-, and IA-administered rats in experiment 3 (hematoxylin and eosin). Moderate atypical hyperplasia featuring an increased number of glands under the lining epithelium (**e,f**). Well-differentiated endometrial adenocarcinoma (AdCa) (**g,h**). Atypical glands present in the endometrium proliferating irregularly and invading the muscle layer (arrows). Stromal polyp (**i,g**) and adenomatous polyp (**k,l**); (**B**) histopathological changes in the uteri of the DMBA, *N*-ethyl-*N'*-nitro-*N*-nitrosoguanidine (ENNG) control (**a,b**), 0.03 (**c,d**), 0.3 (**e,f**), and 1% (**g,h**) PM-treated Donryu rats after initiation; vehicle control (**i,j**) and vehicle 1% PM (**k,l**)-administered rats in experiment 3. Note the development of atypical hyperplasia (DMBA, ENNG → 0.3% PM group), uterus dilatation (0.3 and 1% PM-treated rats), inflammation and hemorrhage (DMBA, ENNG → 0.3 and 1% PM groups) induced by the PM treatment. (Magnifications in (**A**) and (**B**): ×20 (**a,c,e,g,i,k**) and ×200 (**b,d,f,h,j,l**)).

Table 1. Incidence and multiplicity of neoplastic lesions in the mammary glands and uteri of Donryu rats.

Mammary Glands

Treatment	No. Rats [a]	Fibroadenoma	Fibroma	Adenoma	AdCa
		Incidence (No. Rats (%))			
DMBA, ENNG	21	20 (95.2)	7 (33.3)	3 (14.3)	1 (4.8)
DMBA, ENNG → PM, 0.03%	20	19 (95)	8 (40)	2 (10)	6 (30)
DMBA, ENNG → PM, 0.3%	20	19 (95)	8 (40)	4 (20)	7 (35) *
DMBA, ENNG → PM 1%	21	20 (95.2)	5 (23.8)	4 (19.1)	6 (28.6)
DMBA, ENNG → IA, 0.2%	21	18 (85.7)	3 (14.3)	2 (9.5)	6 (28.6)
Vehicle	5	1 (20)	0	0	0
Vehicle → PM, 1%	6	1 (16.7)	0	0	0
		Multiplicity (No./Rat)			
DMBA, ENNG	21	10.90 ± 4.93 [d]	0.55 ± 0.89 [b]	0.15 ± 0.37	0.05 ± 0.22
DMBA, ENNG → PM, 0.03%	20	9.85 ± 5.39	0.40 ± 0.50	0.10 ± 0.31	0.40 ± 0.68
DMBA, ENNG → PM, 0.3%	20	8.30 ± 4.93	0.50 ± 0.69	0.20 ± 0.41	0.45 ± 0.69 [#]
DMBA, ENNG → PM 1%	21	7.95 ± 4.26	0.29 ± 0.56	0.19 ± 0.40	0.33 ± 0.58
DMBA, ENNG → IA, 0.2%	21	6.90 ± 5.02 *	0.14 ± 0.36	0.10 ± 0.30	0.43 ± 0.75
Vehicle	5	0.20 ± 0.45	0	0	0
Vehicle → PM, 1%	6	0.17 ± 0.41	0	0	0

Uterus

Treatment	No. Rats [a]	Dilatation	Endometrial HPL			AdCa	Polyps		
			Mild	Moderate	Severe		Total	S	A
		Incidence (No. Rats (%))							
DMBA, ENNG	21	8 (38.1)	10 (47.6)	5 (23.8)	2 (9.5)	1 (4.8)	17 (81.0)	6 (28.6)	3 (14.3)
DMBA, ENNG → PM, 0.03%	20	13 (65)	16 (80)	4 (20)	1 (5)	3 (15)	17 (85)	5 (25)	2 (10)
DMBA, ENNG → PM, 0.3%	20	19 (95) **	17 (85)	7 (35)	1 (5)	2 (10)	19 (95)	11 (55)	8 (40)
DMBA, ENNG → PM 1%	21	19 (90.5) **	11 (52.4)	2 (9.5)	2 (9.53)	0 (0)	12 (57.1)	4 (19.0)	5 (23.8)
DMBA, ENNG → IA, 0.2%	21	19 (90.5) **	16 (76.2)	5 (23.8)	4 (19.0)	2 (9.5)	20 (95.2)	9 (42.9)	9 (42.9)
Vehicle	5	0 (0)	0 (0)	0 (0)	0 (0)	0 (0)	0 (0)	0 (0)	0 (0)
Vehicle → PM, 1%	6	6 (100) *	3 (50.0)	0 (0)	0 (0)	0 (0)	3 (50.0)	0 (0)	0 (0)
		Multiplicity (No./Rat)							
DMBA, ENNG	21	-	0.67 ± 0.80 [c]	0.24 ± 0.44	0.14 ± 0.48	0.05 ± 0.22	1.10 ± 0.64 [d]	0.33 ± 0.58	0.14 ± 0.36
DMBA, ENNG → PM, 0.03%	20	-	1.25 ± 0.91 *	0.24 ± 0.54	0.05 ± 0.22	0.15 ± 0.37	1.55 ± 1.00	0.25 ± 0.44	0.15 ± 0.47
DMBA, ENNG → PM, 0.3%	20	-	1.35 ± 0.81 *	0.35 ± 0.49	0.05 ± 0.22	0.10 ± 0.31	1.75 ± 0.79 *	0.95 ± 1.19	0.65 ± 1.09
DMBA, ENNG → PM 1%	21	-	0.57 ± 0.60	0.10 ± 0.30	0.14 ± 0.48	0.00 ± 0.00	0.81 ± 0.81	0.43 ± 1.33	0.33 ± 0.73
DMBA, ENNG → IA, 0.2%	21	-	1.00 ± 0.71	0.24 ± 0.44	0.19 ± 0.40	0.10 ± 0.30	1.52 ± 0.60 *	0.57 ± 0.75	0.57 ± 0.75 *
Vehicle	5	-	0	0	0	0	0	0	0
Vehicle → PM, 1%	6	-	0.50 ± 0.55	0	0	0	0.50 ± 0.55	0	0

Values are mean ± SD. [a] Effective number of rats; * Significantly different from the DMBA, ENNG control group at $p < 0.05$; [#] $p = 0.05$; [b–d] Significantly different from the Vehicle control group at $p < 0.05$, $p < 0.01$ and $p < 0.0001$; AdCa, adenocarcinoma; HPL, hyperplasia; S, stromal polyp; A, adenomatous polyp.

Results of histopathological analysis demonstrated significant increase of incidence ($p < 0.05$) and the strong trend for increase of multiplicity of mammary adenocarcinomas in 0.3% PM- ($p = 0.05$) and IA-administered rats (Figure 3C and Table 1). There was a significant negative linear trend for multiplicity of fibroadenoma in PM-treated rats, and significant inhibition in IA-treated rats ($p < 0.05$) (Figure 3D and Table 1).

Macroscopically measured mammary adenocarcinoma volumes were elevated in the 0.03, 0.3, and 1% PM and 0.2% IA groups starting from week 16 as compared to the initiation control rats, with the highest value observed in the 0.03% PM dose group and the lowest increase induced by PM at a dose of 1% (Figure 3E). Unexpectedly, sudden development of benign tumors in the mammary glands of the initiation control rats was observed at weeks 32–36. Animals with higher body weights had larger mammary tumors (Figure 3F).

2.3.4. Histopathological Analysis of Uteri

Data from histopathological examination of rat uteri are shown in Table 1 and Figure 4A(e–l),B(a–l). At termination, the uteri of 0.3 and 1% PM- as well as IA-treated rats after initiation of endometrial carcinogenesis demonstrated dilatation, increased hemorrhage, and higher numbers of nodules macroscopically. In the uteri of N-ethyl-N'-nitro-N-nitrosoguanidine (ENNG)-initiated rats, we observed various proliferative lesions, with a sequence of changes from atypical hyperplasias to adenocarcinomas (Figure 4A(e–h). In addition, stromal and adenomatous polyps were apparent (Figure 4A(i–l). A significant increase in the multiplicity of total atypical hyperplasias (HPLs) (mild, medium, and severe) was found in 0.3% PM- and IA-administered rats (Table 1). Furthermore, significant elevation of multiplicity of mild atypical HPLs was detected in the uteri of 0.03 and 0.3% PM groups. In addition, multiplicities of stromal and adenomatous polyps tended to increase with the 0.3% and 1% PM treatment, while significant elevation was detected in the uteri of the IA-administered rats (Table 1).

2.3.5. Blood Hematology and Biochemistry

The results of hematological and biochemical examinations of the blood are presented in Supplementary Materials Table S2.

Significantly decreased red blood cell count, Hb, and Ht, but increased platelet count, were observed in DMBA- and ENNG-initiated rats as compared to the vehicle control rats. When PM was administered at a dose of 1% without initiation, Ht and lymphocyte count were significantly suppressed but neutrophils elevated, indicating higher levels of inflammation. Furthermore, PM administration after the initiation of mammary and uterine carcinogenesis significantly and dose-dependently suppressed the platelet counts.

In blood biochemistry, significant inhibition and a trend for decrease in total cholesterol and triglyceride levels were found in PM- and IA-administered rats after initiation and in the vehicle 1% PM group. Moreover, in 1% PM- and IA-treated rats, aspartate aminotransferase (AST) and alanine aminotransferase (ALT) were suppressed, whereas alkaline phosphatase (ALP) and γ-glutamyl transpeptidase (γ-GTP) were elevated. Furthermore, in the initiation control and 1% PM vehicle groups, the blood urea nitrogen (BUN) level was increased as compared to the vehicle control group. In addition, significant reduction and trends for decrease of blood calcium levels were found in 1% PM and other PM groups, respectively, as compared to the controls.

3. Discussion

The present results demonstrated that long-term postpubertal exposure to PM at doses higher than 200 mg/kg b.w./day exerts estrogenic activity and induces cell proliferation in the mammary glands of Donryu rats. Furthermore, long-term treatment with PM at 200 mg/kg b.w./day promoted mammary and endometrial carcinogenesis after DMBA and ENNG initiation. Mammary adenocarcinomas metastasizing to the lungs were found in 0.03 and 0.3% PM-treated rats, as well as 0.2% IA treated rats. In this study, the modifying effects of 0.3% PM on mammary gland and uterus were comparable with those of 0.2% IA. In the medium and especially in the high dose PM groups, decreases of rat body weights, adipose deposition, and total cholesterol and triglyceride levels in the blood were obvious, likely due to antilipogenic effects of estrogenic compounds described previously, or perhaps to the decrease of water intake of rats given PM [16]. In the long-term experiment, we observed significant increase of mammary adenocarcinoma development induced by PM at a dose of 0.3%. The absence of dose-dependence in the effects of PM on mammary adenocarcinomas may be related to side effects exerted at high doses, with best incorporation of its ingredients reaching working "physiological" intracellular concentrations at a dose of 0.3%.

Interestingly, short- and long-term administration of PM applied at medium and high doses resulted in increase of uterus weight, and dilatation, hemorrhage, and inflammation of the uterine wall. However, the influence of PM on lesion development in the uterus was much less pronounced as compared to that in the mammary gland—and only atypical hyperplasia was elevated—at 0.3% PM.

Trophic effects of estrogenic compounds on the mammary gland and uterus were previously suggested to be due to activation of signaling through estrogen receptors (ERs) ERα and ERβ [17]. It was reported that PM phytoestrogens at high doses could effectively outcompete 17 β-estradiol binding to ERα in MCF-7 cells [3]. We have previously shown that IA at an estrogenic dose of 150 mg/kg b.w./day activated ERα or ERβ and downstream AP1 and NF-κB transcriptional factors, also potentiating F-actin signaling in mammary and uterine adenocarcinomas [15]. However, the effects of biological substances possessing estrogenic activity generally appear to be dependent on the dose. In case of IA intake, "physiological" concentrations are known as those achieved in the serum of persons consuming commonly recommended daily doses of isoflavones of 50–100 mg [18]. However, in the case of PM, there are almost no data concerning the concentration of ingredients in the blood and tissues. The safe dose for humans is considered as 1–2 mg/kg b.w./day. From our results, the dose of PM exerting promoting effects on mammary and uterine carcinogenesis in rats was close to 200 mg/kg b.w./day (0.3%).

In female monkeys, daily treatment with 100 mg and 1000 mg/day (about 20 and 200 mg/kg b.w./day) of PM for 90 days produced a dose-dependent reduction in the urinary follicular stimulating hormone (FSH), luteinizing hormone (LH) and estradiol levels in the blood, and the single dose of 1000 mg disturbed ovarian function and menstrual cycling [19–22]. Furthermore, recent experiments in mice demonstrated that oral exposure to a nontoxic PM dose of 100 mg/kg b.w./day for 8 weeks resulted in prolonged estrous cycles, while 10 mg/kg b.w./day did not induce any changes in the hypothalamic–pituitary–ovarian–uterine axis, and did not exert estrogenic activity or adverse effects on mating efficiency or reproduction [23]. In addition, development of uterine endometrial hyperplasia and a decrease in the number of growing ovarian follicles, possibly related to reduction in the luteinizing hormone (LH) and follicle stimulating hormone (FSH) levels, were detected after PM application to mice at a dose of 100 mg/kg b.w./day but not at 10 mg/kg b.w./day. Moreover, in studies with gonadectomized female rats, oral treatment with water-suspended PM at doses of 100 and 1000 mg/kg b.w./day for 2 weeks resulted in a significant increase of uterine weight, remarkable vaginal and uterine proliferation, vaginal cornification, and suppression of reproductive functions [24–28]. Recovery after cessation of treatment was dependent on the dosage of PM.

In line with the previous results in rats and mice, in our short-term study we observed estrogenic effects of PM applied to ovariectomized rats at doses of 200 and 2000 mg/kg b.w./day for 2 weeks. Long-term PM administration at a dose of 0.3% (200 mg/kg b.w./day) was found not only to exert estrogenic activity in the mammary gland and uterus, but also to promote mammary carcinogenesis and induce atypical hyperplasia in the uteri of Donryu rats.

It is important to further mention that the timing of exposure to substances with estrogenic bioactivities is thought to be critical for effects on breast cancer risk. Thus, prepubertal and postpubertal exposure to estrogenic compounds such as genistein could have different effects on cell proliferation in the terminal ductules of mammary glands [29]. In the present case, postpubertal exposure to

PM, isoflavones or other test compounds with estrogenic activity induced cell proliferation and promoted mammary and uterine carcinogenesis in our two-step carcinogenesis model with DMBA and ENNG initiation.

We also observed that 1% PM suppressed lymphocyte and platelet counts but elevated neutrophil levels, presumably reflecting effects on the immune system, promoting inflammation, and bone marrow suppression. Moreover, 1% PM elevated the ALP and γ-GTP as well as the BUN in the blood, which could occur if kidneys or liver were damaged. In addition, the present results demonstrated that 1% PM induced a decrease in blood calcium levels. These data are in line with recent results demonstrating that long-term treatment of aged menopausal monkeys with 1000 mg/day of PM decreased serum parathyroid hormone (PTH) and calcium levels in the blood likely due to the amelioration of the bone loss caused by estrogen deficiency [11,12].

In conclusion, in the present study long-term postpubertal treatment of Donryu rats with PM at a dose of 200 mg/kg b.w./day exerted promoting effects on mammary carcinogenesis after the initiation with DMBA. Furthermore, PM elevated cell proliferation in the mammary glands of DMBA-initiated rats, which might lead to the promotion and progression of mammary tumors to greater malignancy. In addition, it inhibited the levels of calcium in the blood, and induced inflammation, hemorrhage, and dilatation of the uterine wall in rats.

4. Experimental Section

4.1. Chemicals

N-ethyl-N'-nitro-N-nitrosoguanidine (ENNG) was obtained from Nakalai Tesque (Kyoto, Japan), 7,12-dimethylbenz[a]anthracene (DMBA) from Tokyo Chemical Industry Co. Ltd. (Tokyo, Japan) and polyethylene glycol (PEG) from Wako Pharmaceutical, Osaka, Japan. Other chemicals were from Sigma Chemical Co. (St Louis, MO, USA) or Wako Pharmaceutical (Osaka, Japan).

4.2. Test Compounds

The *Pueraria mirifica* powder (Lot No.: PM490621) was produced by Seiko Yakuhin Kogyo K.K. (Narashino, Chiba, Japan) and consigned by Shiratori Pharmaceutical Co., Ltd., (Narashino, Chiba, Japan) and Pias Corporation (Osaka, Japan). The taxonomic and content identification was performed by the Seiko Yakuhin Kogyo K.K. The sample contained miroestrol (5.3 mg/kg; 0.00053%), deoxymiroestrol (6.3 mg/kg; 0.00063%), puerarin (6'-O-beta-apiofuranoside: 21.7 mg/kg; 0.00217%), daidzin (daidzein-7-O-glucoside: 12.9 mg/kg; 0.00129%), genistin (genistein-7-O-glucoside: 8.7 mg/kg; 0.00087%), daidzein (7,4'-dihydroxyisoflavone: 48.2 mg/kg; 0.00482%), genistein (25.5 mg/kg; 0.00255%) and kwakhurin (3-[2-(3,3-dimethylallyl)-4,6-dihydroxy-3-methoxyphenyll-7-hydroxyisoflavone: 3.5 mg/kg; 0.00035%).

Isoflavone aglycon (IA) extract (SoyAct) was from Kikkoman Corporation (Noda City, Chiba, Japan). In the present experiment, test powder diets were prepared as follows: 0.03%, 0.3%, 1, and 3% PM diets contained 0.03%, 0.3%, 1, and 3% *Pueraria mirifica* powder, respectively, in NIH-07PLD powder diet (Phytoestrogen Low Diet, Oriental Yeast, Tokyo, Japan). The accuracy of dose formulation and uniformity of blending of the diets was confirmed by the analytical chemistry laboratories at Oriental Yeast Co., Tokyo, Japan. For the production of IA extract fermentation of soy was performed followed by ethanol/water extraction and purification [15].

4.3. Animals

One hundred and seventeen female 4-week-old Crlj:DON (Donryu) rats (Japan SLC, Shizuoka, Japan) were obtained at 5 weeks of age and quarantined for 1 week before the experiment started. The rats were kept in an animal house with a 12 h (8:00–20:00) light/dark cycle, humidity of 50% ± 2% and temperature of 23 ± 1 °C. Tap water and NIH-07PLD diet (Phytoestrogen Low Diet, Oriental Yeast, Tokyo, Japan) was given ad libitum. NIH-07PLD diet constituents were carbohydrate, crude protein, crude fiber, fat, neutral detergent fiber, ash, fatty acids, amino acids, vitamins, and trace elements with no phytoestrogens. All animals were checked once daily for general behavior and signs of toxicity or a moribund state. Body weights, water and food intakes were measured every week for the first 12 weeks, and thereafter every 4 weeks. During the experiment, the specific signs used to determine when the animals should be euthanized included no response to stimuli or a comatose condition, changes in heart rate and physical appearance, dyspnea or severe breathing problems, hypothermia, prostration, body weight loss, and changes in food and water intakes. If significant body weight loss or the water and food consumption changes were detected, the relevant animals were checked more precisely for other signs of sickness. At euthanization, a systemic macroscopic pathological examination of liver, kidneys, spleen, adrenals, thymus, mammary glands, and uterus was performed. All experimental procedures were conducted with approval and according to the Guidelines of Animal Care and Use Committee of the Osaka City University Medical School.

4.4. Short-Term Experiment 1

We performed ovariectomies on 25 female Donryu rats (5 weeks of age) with normal estrous cycles. All animals were checked by vaginal cytology to confirm the absence of estrous cycles. Two weeks after the ovariectomy, for 2 weeks the rats were given (1) PM at doses of 0.03%, 0.3%, and 3%; (2) 0.2% isoflavone aglycone (IA) extract in basal NIH-07PLD diet; or (3) only control diet. Vaginal smears stained with aqueous solution of methylene blue were used to check the estrogenic activity of the test compounds. Animal body weights were measured once a week, and general condition was examined once a day. The weights of uteri were measured at final necropsy to determine estrogenic effects of test compounds. Mammary gland, uterus, liver, kidneys, spleen, adrenals, and thymus were subjected to histopathological analysis.

Donryu rats with a mean body weight of 200 g consumed PM in 15 g diet at doses of 20 (0.03%), 200 (0.3%), 667 (1%), and 2000 (3%) mg/kg b.w./day, considered equal to about 0.2, 2.0, and 20.0 mg/kg b.w./day (10, 50, and 1000 mg/day) intake by women with mean body weight of 50 kg (the acceptable daily intake (ADI) for rats is considered to be 100 times that of humans as the safety factor is 100 (World Health Organization)). As the ADI of PM for humans is 1–2 mg/kg b.w./day, with this extrapolation, the 0.3% PM dose used in our experiment would be accepted as safe for humans (2 mg/kg b.w./day). In addition, the dose for rats could be also extrapolated to a human equivalent dose by the body surface area (BSA) normalization method (mg/m^2 conversion) [30,31], in which multiplication of the human dose by 6.16 is applied (Km human/Km animal = 37/6). In case of BSA normalization, PM doses of 20, 200, and 2000 mg/kg b.w./day would be equal to 3.2, 32.5, and 325 mg/kg b.w./day intake in humans, and the accepted safe dose of PM is close to 0.03%.

Because of the very strong estrogenic effect of 3% PM observed in this experiment, the largest dose applied in the long-term experiment was changed to 1% (667 mg/kg b.w./day), equal to about 7 mg/kg b.w./day (350 mg/day) for humans, or in the case of BSA normalization, 108 mg/kg b.w./day intake.

4.5. Short-Term Experiment 2

DMBA in sesame oil at a dose of 50 mg/kg b.w. was administered by gavage (i.g.) to 20 female Donryu rats (5 weeks of age). Another 5 rats were given an equivalent volume of sesame oil alone (~0.5 mL/rat). PM at doses of 0.03%, 0.3%, and 3% and 0.2% IA were administered in NIH-07PLD diet for 4 weeks starting from the day of DMBA injection. At euthanasia, liver, kidneys, spleen, adrenals, thymus, and uterus were weighed, and samples of mammary glands were fixed in 10% phosphate-buffered formalin for the histopathological and bromodeoxyuridine (BrdU) immunohistochemical examination.

4.6. Long-Term Experiment (Experiment 3)

One hundred and seventeen 5-week-old female Donryu rats were divided into seven groups. It is known that soon after weaning, about postnatal day 35, pubertal development of Donryu rats starts. Since the onset of puberty is defined as the age (in days) at which vaginal opening occurs, rats were inspected daily for this purpose. At the commencement of the experiment, 5-week-old rats from PM-treated, IA-treated, and initiation control groups (21 rats/group) were given a single dose of DMBA by i.g. (50 mg/kg b.w.) for initiation of mammary carcinogenesis. On experimental days 7 and 11, ENNG (10 mg/kg b.w.) in PEG was injected via the vagina using a stainless catheter to initiate uterine carcinogenesis. Six rats each in vehicle and 1% PM vehicle groups received sesame oil by i.g. and PEG via the vagina. PM was administered to rats at doses of 0.03%, 0.3%, and 1%, and IA was applied for comparison at a dose of 0.2% in NIH-07PLD basal diet for 36 weeks from the commencement of the experiment. Rats in the vehicle control group received the basal diet. One of the characteristics of Donryu rats is age-related persistent estrus followed by anovulation starting at the age of 5 months, the incidence of which rises until 8 months [32]. Mammary tumor number and volume (cm^3/rat; the formula: tumor length/2 $\times$ (width/2)^2) were assessed once weekly, and the location of each nodule was recorded. Malignant tumors usually metastasized to the lung, contained abscesses and ulcers, and were of dark color. Histopathologic analysis was performed according to the previously published classification of mammary tumors [33].

All surviving animals were euthanized at week 36 and the test organs—including mammary glands with skin, uterus, vaginas, ovaries, liver, pituitary, adrenals, and thymus—were removed, weighed, and fixed in 10% formalin for histopathological analysis. Twelve specimens were obtained from each uterus in cross-section and proliferative endometrial lesions were classified using the categories reported previously [34] into three degrees (mild, moderate, and severe) of atypical hyperplasia, stromal and adenomatous polyps, and adenocarcinoma.

4.7. BrdU Immunohistochemistry

In short-term experiment 2, for evaluation of cellular proliferation, BrdU staining was performed for rat mammary glands by the avidin-biotin-peroxidase complex (ABC) method reported previously [35]. Sections were incubated with mouse monoclonal anti-BrdU antibody (Dako Japan, Kyoto, Japan) at 1:500 dilution. Immunoreactivity was detected using a Vectastain Elite ABC Kit (PK-6102; Vector Laboratories, Burlingame, CA, USA) and 3,3′-diaminobenzidine hydrochloride (Sigma Chemical Co., St. Louis, MO, USA). A negative control was also included with the staining procedure by omitting the primary antibody. At least 3000 mammary epithelial cells' nuclei were counted in each rat, and labeling indices were calculated as numbers of positive nuclei per 1000 cells.

4.8. Blood Hematology and Biochemistry

Blood samples were collected directly from the hearts of all surviving rats at the end of the study after overnight fasting. An automated hematology analyzer (Sysmex XE-2100, Mitsubishi Chemical Visuals, Osaka, Japan) and an automatic analyzer (Olympus AJ-5200, Tokyo, Japan) were employed for hematological and biochemical analyses of blood serum as previously described [36].

4.9. Statistical Analysis

Statistical analysis was performed using the StatLight-2000(C) program (Yukms Corp., Tokyo, Japan) or with GraphPad Prism 5 Software Inc. (San Diego, CA, USA). Incidences of histopathological lesions were compared by the Fisher's exact probability test or the χ^2-test, and the log rank test. The Kaplan–Meier method was employed for assessment of differences in survival. Numerical data for control and experimental groups were statistically compared using the Bartlett's test. The Dunnett's multiple comparison test (two-sided) was used in case of homogeneous data, otherwise the Steel's test (two-sided) was applied [37]. For all data, P values less than 0.05 were considered significant.

Supplementary Materials: The following are available online at www.mdpi.com/2072-6651/8/11/275/s1, Table S1: Final body, relative organ weights and chemicals intakes of Donryu rats in Exp. 3, Table S2: Results of the hematological and blood biochemical analyses.

Acknowledgments: This work was supported by a Grant-in-Aid for Scientific Research from the Ministry of Health, Labor and Welfare of Japan.

Author Contributions: A.K. and H.W. conceived and designed the experiments; A.K., M.Y. and Y.T. performed the experiments and analyzed the data; M.Y. contributed reagents/materials/analysis tools; A.K., N.I., T.O. and M.G. wrote the paper.

Conflicts of Interest: The authors declare no conflict of interest.

References

1. Subcharoen, P. Thai traditional medicine in the new millennium. *J. Med. Assoc. Thail.* **2004**, *87* (Suppl. S4), S52–S57.

2. Ingham, J.L.; Tabara, S.; Pope, G.S. *Chemical Components and Pharmacology of the Rejuvenating Plant Pueraria mirifica*; Taylor and Fransis: London, UK, 2002.

3. Malaivijitnond, S. Medical applications of phytoestrogens from the Thai herb *Pueraria mirifica*. *Front. Med.* **2012**, *6*, 8–21. [CrossRef] [PubMed]

4. Cain, J.C. Miroestrol: An oestrogen from the plant *Pueraria mirifica*. *Nature* **1960**, *188*, 774–777. [CrossRef] [PubMed]

5. Shimokawa, S.; Kumamoto, T.; Ishikawa, T.; Takashi, M.; Higuchi, Y.; Chaichantipyuth, C.; Chansakaow, S. Quantitative analysis of miroestrol and kwakhurin for standardisation of Thai miracle herb 'Kwao Keur' (*Pueraria mirifica*) and establishment of simple isolation procedure for highly estrogenic miroestrol and deoxymiroestrol. *Nat. Prod. Res.* **2013**, *27*, 371–378. [CrossRef] [PubMed]

6. Chansakaow, S.; Ishikawa, T.; Seki, H.; Sekine, K.; Okada, M.; Chaichantipyuth, C. Identification of deoxymiroestrol as the actual rejuvenating principle of "Kwao Keur", *Pueraria mirifica*. The known miroestrol may be an artifact. *J. Nat. Prod.* **2000**, *63*, 173–175. [CrossRef] [PubMed]

7. Pope, G.S.; Grundy, H.M.; Jones, H.E.H.; Tait, S.A.S. The oestrogenic substance (miroestrol) from the tuberous roots of *Pueraria mirifica*. *J. Endocrinol.* **1958**, *17*, XV–XVI.

8. Chansakaow, S.; Ishikawa, T.; Sekine, K.; Okada, M.; Higuchi, Y.; Kudo, M.; Chaichantipyuth, C. Isoflavonoids from *Pueraria mirifica* and their estrogenic activity. *Planta Med.* **2000**, *66*, 572–575. [CrossRef] [PubMed]

9. Tiyasatkulkovit, W.; Charoenphandhu, N.; Wongdee, K.; Thongbunchoo, J.; Krishnamra, N.; Malaivijitnond, S. Upregulation of osteoblastic differentiation marker mRNA expression in osteoblast-like UMR106 cells by puerarin and phytoestrogens from *Pueraria mirifica*. *Phytomedicine* **2012**, *19*, 1147–1155. [CrossRef] [PubMed]

10. Tiyasatkulkovit, W.; Malaivijitnond, S.; Charoenphandhu, N.; Havill, L.M.; Ford, A.L.; VandeBerg, J.L. *Pueraria mirifica* extract and puerarin enhance proliferation and expression of alkaline phosphatase and type I collagen in primary baboon osteoblasts. *Phytomedicine* **2014**, *21*, 1498–1503. [CrossRef] [PubMed]

11. Suthon, S.; Jaroenporn, S.; Charoenphandhu, N.; Suntornsaratoon, P.; Malaivijitnond, S. Anti-osteoporotic effects of *Pueraria candollei* var. mirifica on bone mineral density and histomorphometry in estrogen-deficient rats. *J. Nat. Med.* **2016**, *70*, 225–233. [CrossRef] [PubMed]

12. Trisomboon, H.; Malaivijitnond, S.; Suzuki, J.; Hamada, Y.; Watanabe, G.; Taya, K. Long-term treatment effects of *Pueraria mirifica* phytoestrogens on parathyroid hormone and calcium levels in aged menopausal cynomolgus monkeys. *J. Reprod. Dev.* **2004**, *50*, 639–645. [CrossRef] [PubMed]

13. Cherdshewasart, W.; Panriansaen, R.; Picha, P. Pretreatment with phytoestrogen-rich plant decreases breast tumor incidence and exhibits lower profile of mammary ERalpha and ERbeta. *Maturitas* **2007**, *58*, 174–181. [CrossRef] [PubMed]

14. Lucki, N.C.; Sewer, M.B. Genistein stimulates MCF-7 breast cancer cell growth by inducing acid ceramidase (ASAH1) gene expression. *J. Biol. Chem.* **2011**, *286*, 19399–19409. [CrossRef] [PubMed]

15. Kakehashi, A.; Tago, Y.; Yoshida, M.; Sokuza, Y.; Wei, M.; Fukushima, S.; Wanibuchi, H. Hormonally active doses of isoflavone aglycones promote mammary and endometrial carcinogenesis and alter the molecular tumor environment in Donryu rats. *Toxicol. Sci.* **2012**, *126*, 39–51. [CrossRef] [PubMed]

16. Santana, A.C.; Soares da Costa, C.A.; Armada, L.; de Paula Lopes Gonzalez, G.; dos Santos Ribeiro, M.; de Sousa dos Santos, A.; de Carvalho, J.J.; do Nascimento Saba, C.C. Fat tissue morphology of long-term sex steroid deficiency and estrogen treatment in female rats. *Fertil. Steril.* **2011**, *95*, 1478–1481. [CrossRef] [PubMed]

17. Messina, M.J.; Wood, C.E. Soy isoflavones, estrogen therapy, and breast cancer risk: Analysis and commentary. *Nutr. J.* **2008**, *7*, 17. [CrossRef] [PubMed]

18. Wuttke, W.; Jarry, H.; Seidlova-Wuttke, D. Isoflavones-safe food additives or dangerous drugs? *Ageing Res. Rev.* **2007**, *6*, 150–188. [CrossRef] [PubMed]

19. Trisomboon, H.; Malaivijitnond, S.; Watanabe, G.; Taya, K. Estrogenic effects of *Pueraria mirifica* on the menstrual cycle and hormone-related ovarian functions in cyclic female cynomolgus monkeys. *J. Pharmacol. Sci.* **2004**, *94*, 51–59. [CrossRef] [PubMed]

20. Trisomboon, H.; Malaivijitnond, S.; Watanabe, G.; Taya, K. Ovulation block by *Pueraria mirifica*: A study of its endocrinological effect in female monkeys. *Endocrine* **2005**, *26*, 33–39. [CrossRef]

21. Trisomboon, H.; Malaivijitnond, S.; Watanabe, G.; Cherdshewasart, W.; Taya, K. The estrogenic effect of *Pueraria mirifica* on gonadotrophin levels in aged monkeys. *Endocrine* **2006**, *29*, 129–134. [CrossRef]

22. Trisomboon, H.; Malaivijitnond, S.; Cherdshewasart, W.; Watanabe, G.; Taya, K. Assessment of urinary gonadotropin and steroid hormone profiles of female cynomolgus monkeys after treatment with *Pueraria mirifica*. *J. Reprod. Dev.* **2007**, *53*, 395–403. [CrossRef] [PubMed]

23. Jaroenporn, S.; Malaivijitnond, S.; Wattanasirmkit, K.; Watanabe, G.; Taya, K.; Cherdshewasart, W. Assessment of fertility and reproductive toxicity in adult female mice after long-term exposure to *Pueraria mirifica* herb. *J. Reprod. Dev.* **2007**, *53*, 995–1005. [CrossRef] [PubMed]

24. Malaivijitnond, S.; Kiatthaipipat, P.; Cherdshewasart, W.; Watanabe, G.; Taya, K. Different effects of *Pueraria mirifica*, a herb containing phytoestrogens, on LH and FSH secretion in gonadectomized female and male rats. *J. Pharmacol. Sci.* **2004**, *96*, 428–435. [CrossRef] [PubMed]

25. Malaivijitnond, S.; Chansri, K.; Kijkuokul, P.; Urasopon, N.; Cherdshewasart, W. Using vaginal cytology to assess the estrogenic activity of phytoestrogen-rich herb. *J. Ethnopharmacol.* **2006**, *107*, 354–360. [CrossRef] [PubMed]

26. Cherdshewasart, W.; Kitsamai, Y.; Malaivijitnond, S. Evaluation of the estrogenic activity of the wild *Pueraria mirifica* by vaginal cornification assay. *J. Reprod. Dev.* **2007**, *53*, 385–393. [CrossRef] [PubMed]

27. Cherdshewasart, W.; Sriwatcharakul, S.; Malaivijitnond, S. Variance of estrogenic activity of the phytoestrogen-rich plant. *Maturitas* **2008**, *61*, 350–357. [CrossRef] [PubMed]

28. Gomuttapong, S.; Pewphong, R.; Choeisiri, S.; Jaroenporn, S.; Malaivijitnond, S. Testing of the estrogenic activity and toxicity of Stephania *venosa* herb in ovariectomized rats. *Toxicol. Mech. Methods* **2012**, *22*, 445–457. [CrossRef] [PubMed]

29. Lamartiniere, C.A.; Moore, J.B.; Brown, N.M.; Thompson, R.; Hardin, M.J.; Barnes, S. Genistein suppresses mammary cancer in rats. *Carcinogenesis* **1995**, *16*, 2833–2840. [CrossRef] [PubMed]

30. Freireich, E.J.; Gehan, E.A.; Rall, D.P.; Schmidt, L.H.; Skipper, H.E. Quantitative comparison of toxicity of anticancer agents in mouse, rat, hamster, dog, monkey, and man. *Cancer Chemother. Rep.* **1966**, *50*, 219–244. [PubMed]

31. Reagan-Shaw, S.; Nihal, M.; Ahmad, N. Dose translation from animal to human studies revisited. *FASEB J.* **2008**, *22*, 659–661. [CrossRef] [PubMed]

32. Nagaoka, T.; Takegawa, K.; Takeuchi, M.; Maekawa, A. Effects of reproduction on spontaneous development of endometrial adenocarcinomas and mammary tumors in Donryu rats. *Jpn. J. Cancer Res.* **2000**, *91*, 375–382. [CrossRef] [PubMed]

33. Costa, I.; Solanas, M.; Escrich, E. Histopathologic characterization of mammary neoplastic lesions induced with 7,12 dimethylbenz(alpha)anthracene in the rat: A comparative analysis with human breast tumors. *Arch. Pathol. Lab. Med.* **2002**, *126*, 915–927. [PubMed]

34. Nagaoka, T.; Takeuchi, M.; Onodera, H.; Matsushima, Y.; Ando-Lu, J.; Maekawa, A. Sequential observation of spontaneous endometrial adenocarcinoma development in Donryu rats. *Toxicol. Pathol.* **1994**, *22*, 261–269. [CrossRef] [PubMed]

35. Kakehashi, A.; Hagiwara, A.; Imai, N.; Wei, M.; Fukushima, S.; Wanibuchi, H. Induction of cell proliferation in the rat liver by the short-term administration of ethyl tertiary-butyl ether. *J. Toxicol. Pathol.* **2015**, *28*, 27–32. [CrossRef] [PubMed]

36. Kakehashi, A.; Ishii, N.; Fujioka, M.; Doi, K.; Gi, M.; Wanibuchi, H. Ethanol-extracted Brazilian propolis exerts protective effects on tumorigenesis in Wistar Hannover rats. *PLoS ONE* **2016**, *11*, e0158654. [CrossRef] [PubMed]
37. Kobayashi, K.; Kanamori, M.; Ohori, K.; Takeuchi, H. A new decision tree method for statistical analysis of quantitative data obtained in toxicity studies on rodent. *Sangyo Eiseigaku Zasshi* **2000**, *42*, 125–129. [PubMed]

MDPI AG

St. Alban-Anlage 66

4052 Basel, Switzerland

Tel. +41 61 683 77 34

Fax +41 61 302 89 18

http://www.mdpi.com

Toxins Editorial Office

E-mail: toxins@mdpi.com

http://www.mdpi.com/journal/toxins

www.ingramcontent.com/pod-product-compliance
Lightning Source LLC
Chambersburg PA
CBHW051906210326
41597CB00033B/6039